MEMORIES OF THE

Northumberland Coalfields

Neil Taylor

COUNTRYSIDE BOOKS
NEWBURY BERKSHIRE

First published 2009
© Neil Taylor, 2009

COUNTRYSIDE BOOKS
3 Catherine Road
Newbury, Berkshire

To view our complete range of books,
please visit us at
www.countrysidebooks.co.uk

ISBN 978 1 84674 100 5

Designed by Peter Davies, Nautilus Design

Produced through MRM Associates Ltd., Reading
Typeset by CJWT Solutions, St Helens
Printed by Information Press, Oxford

*All material for the manufacture of this book
was sourced from sustainable forests.*

Contents

Acknowledgements

To my old mining marras, friends, associates, and their families I wish to give my heartfelt thanks for sharing with me their memories, stories and pictures which make up a substantial part of this publication: Jack Tubby, Kit Miller, Mike Kirkup, Ian Lavery, Alex Anderson, Jim Sawyer, Iris Foster, Jimmy Swan, Doris Teasdale, Alan Stewart, Geoff Murray, Gladys Aynsley, Stan Elliot, Barry Mead, Alan Young, Keith Adams, Eric Dodds, Bob Bell, Ken Hutchinson, Jim Locker, John Freeman, Dave Brown, Tom Fleming, Ron Simpson, Sid Stowe, Phil Renforth, Jos Hanson, George Hume, Owen Taylor, Henry Cleverley, Sheila Graham, Eileen Brown, Mel Brown, Alf Goodall, Bob Charlton, Jack Boaden, Alan Reed, Margaret Nicholson, Bob Snaith.

I would also like to acknowledge my debt to the *Colliery Guardian*; Durham Mining Museum; Great Northern Coalfield; County Records; Ashington Collieries Magazines; Northumberland Miners' Union; Collieries in Northumberland; *Blood On The Coal*; *Their Lesson Our Inspiration*; *Coal News*.

Dedication

A special mention for my dear friend Iris Foster and my former neighbour Alf Goodall, two Lynemouth village stalwarts who sadly will not see the publication of this book. It is folk like them who leave a legacy of life in a pit village through their involvement in the community and who will always be remembered.

Introduction

The miners of Northumberland are proud people, with a culture, heritage and language all their own in England's most northerly county. Their blood still runs black, their dignity, pride and humour are still intact, and this shines through in the many ex-miners and their families I have had the privilege to know and interview in producing this book.

Coal has been mined in the county for almost 2,000 years, when we consider that the Romans building Hadrian's Wall knew of the mineral's existence and left traces of its use. The monks of Holy Island from at least the 6th century were also familiar with fossil fuel where coal seams outcropped along the sea cliffs and sea bed of the East Coast.

Coal was never as popular as charcoal for domestic purposes during the Middle Ages, but in places where it was easily accessible it was used in the smiths' forges.

There is no doubt that the rape of the country's forests up until the 13th century resulted in a shortage of timber that led to the upsurge in the production of coal throughout the land. Northumberland was blessed with coal in abundance, a product that would eventually fuel the industrial revolution of the 18th and 19th centuries and make Britain the world's powerhouse.

In 1239 a special charter was granted by Henry III to the good men of Newcastle to dig coals in the common soil of the ground. The friars of Tynemouth Priory owned and leased a colliery in 1330 and a further record in 1530 shows they leased two mines at a yearly rent of £50.

The coal trade in the county was centred on the Newcastle area of Tyneside with its easy access from the numerous pits that had sprung up along the banks of the River Tyne. Loaded into keel

Newcastle keel boats

boats, then transferred to ships, the coals would be ferried to various ports in the country and especially to London. The ladies of the city were not at all impressed by what they described as the dirty, dusty, smelly sea coal. They called it sea coal because of the way it was transported. It was not until the reign of Edward III that the export of coal from the kingdom was allowed and then only to the port of Calais in France.

By 1615 the number of vessels carrying Newcastle coals had risen to 400 and trade with the French increased with as many as 50 foreign ships leaving the Tyne exporting coal to Picardy, Bretagne and Normandy. Further north the Northumberland ports of Hartley, Blyth and – from the 19th century – Amble became important dispersal points for exporting the black diamonds.

The county in the 18th and early 19th centuries was riddled with shallow pits which became deeper and more sophisticated as machinery was introduced for pumping water and winding coal to the surface. The steam engine and the progress of the railways meant coal could be transported more quickly from the inland mines to the coast. Some of the pits were blessed with an abundance of seggar clay which was mined and used in the building industry. Brickworks were established by the relevant coal companies who built their employees' homes and social outlets, and exported bricks worldwide. Two fine examples were Ashington Coal Company's site at Ashington and the Burn Fireclay Company's Stobswood outlet.

During the 20th century, the old collieries' reserves of coal were being exhausted. Competition from the oil and gas industries increased during the 1950s and 1960s so many of these pits closed. The emphasis was now on the coastal pits with vast reserves under the North Sea, a smattering of viable smaller concerns, and an increase in opencast mining.

After the infamous 1984–5 miners' strike, the Conservative government decimated the coal industry's deep mines, until eventually Ellington Combine became the last colliery in the county to close in January 2005.

There is an old saying in the mining fraternity, 'It's like taking coals to Newcastle', meaning not necessary, a strange thing to do when the coals are there anyway. Ironically this is now the case with imported coal being handled at the ports of Tyne and Blyth and transported off to power stations nationwide. Whatever would the old colliers have thought about that?

Stobswood brickyard was established in 1923 and exported bricks worldwide. It is pictured here in the new millennium shortly before demolition.

The Great Northern Coalfield

The Northumberland Coalfield is part of the triangular-shaped Great Northern Coalfield that stretches from the centre of County Durham up to the north of Northumberland at the Scottish border, and from the Pennine hills in the west across to the east and out under the North Sea.

The section of this coalfield that lies within the confines of Northumberland is divided into three parts by the Great Whinsill Dyke. This is a formidable barrier of hard igneous rock that was created way back in the mists of time when molten rock poured into fissures in the earth's crust. It extends from the north of Berwick across to the Farne Isles and then veers south before turning west to cut across the north side of Newcastle and on into Cumbria where it terminates near Brampton.

After the retreat of the last Ice Age exposed this volcanic rock, it became a natural line of defence against marauding Scots. Roman leader Hadrian built his famous wall along part of its length in the west of the county. The great coastal castles of Northumberland such as Bamburgh, Lindisfarne and Dunstanburgh are all built on its massive outcrop of hard igneous

An early picture of Hebburn Colliery.

rock. An asset to defence of the county it may have been, but to the miners it was nothing more than a massive barrier that separated the coal measures. To the north of the Whinsill are the Scremerston Coal Measures, consisting of six workable seams of normally low quality coal. By 1947 only one pit was producing coal in this area.

In the west are the Tynedale Measures in the limestone strata with twelve workable seams of varying quality. The vast majority of these are in an area around the Cumbrian border. Only three of these seams can be found throughout the whole of the Tynedale Measures. After the Second World War only seven mines worked this area.

The most prolific and productive coal seams are found in the east of the county where they lie between layers of carboniferous limestone and sandstone. There are 24 seams in total, formed much earlier than those in the middle and upper limestone series.

Pitmatic: 'The way wi taak, man'
'Pitmatic' is a language that evolved in the coalfields during the 18th and 19th centuries and combined with the old rural Northumbrian dialect is unique, with many of the words traced way back to the Anglo-Saxon and Viking influence and more than difficult to understand for anyone not from a mining area or background. Sadly, with the demise of the county's pits and demolition of the once thriving mining villages, our language is in decline. In the decades ahead when the last remnants of the old pit regime fade away, true pitmatic will be used less and less by the man in the street. Thankfully, the pitmatic has been recorded in many ways over the years. Tapes, CDs, music and the written word all play a part in ensuring that our rich, earthy pitmatic language will never be lost. Listed here are examples of everyday words and phrases used by pitmen and their families, some of which appear in various chapters in this publication.

Bairn	A child or young person.
Bait	Snack, lunch.
Bank	Surface area at pithead.
Caunch	Stone above roadway entrance to coal face.
Cavil	Choosing workplace by drawing lots.
Chaakin on the bleezer	Not speaking to family member, a rift, leaving written messages.
Champion	Excellent, great.
Chocks	Hardwood billets built up to support roof after coal extraction.
Chummin	Empty tub or mine car.
Clag	To stick or paste, or to hit someone.
Clinkers	Impure coal left after burning.
Cowp	To tip up or to swop something.
Crack	Conversation, chat, news.
Cracket	Low wooden stool as used by hewers when picking at coal face.
Creep	Heaving up of the bottom after coal extraction.
Crowdy	Hen food mixture, grain, bread, potato peelings.
Goaf	Waste area left after coal extraction.
Dad	Father, or a slap or smack.
Deputy	Man in charge of workplace underground.
Drawer	Moves on belt after coal filling draws out worked area.
Ducket	Pigeon loft or shed.
Duff	Small coals and dust left after undercutting coal seam.
Dunch	To bump into.
Dyke	A hard barrier of basalt and other volcanic rocks between coal measures.
Ettled	Ready or getting prepared for work or any occasion.
Femmer	Fragile, weak.
Gallowa	Pony, named from Galloway in Scotland, home of early pit ponies.
Gannin Yem	Going home.
Ganny	Grandmother or an old lady.

Gliff	To frighten or shock or to catch sight of something.
Gowk	A fool or an apple core.
Gully	Large knife for cutting bread or other food.
Howk	To dig or scoop out.
Inbye/ Outbye	In to the coalface and out to the shaft.
Keks	Trousers, pants.
Kep	To catch, as in 'kep the ball'.
Keps	Cage stops.
Kist	Deputy's tool box and meeting place.
Knaa	To know.
Lowse	To loosen, or the end of the working shift.
Mam	Mother.
Marras	Workmates, close friends.
Netty	Outside toilet or any toilet.
Paaky	Choosy over food, or hard to please about many things.
Rolleywayman	Maintains underground rope haulage and rail tracks.
Stowed Off	Overcrowded, full to capacity.
Tek the gee	Act up or take offence.
Tetties	Potatoes.
Timma Leada	Young lad who uses pony to take timber supports to coal face.
Whey Aye	Yes, of course.

When talking to ex-miners and their families, nothing is held back, the folk of the mining communities see and say things as they are, a spade is a spade, nothing more or less. As the years pass, there will be fewer ex-colliers and families to record how it was, better to do it now while the flame burns bright.

The period covered by this book begins in the early 20th century and continues on until the closure of the last pit in the county at Ellington in 2005. Humour, sadness, disaster and resilience are all to be found here. These are their memories and stories, given straight from the heart.

Neil Taylor

MAP OF THE NORTHUMBERLAND COALFIELDS

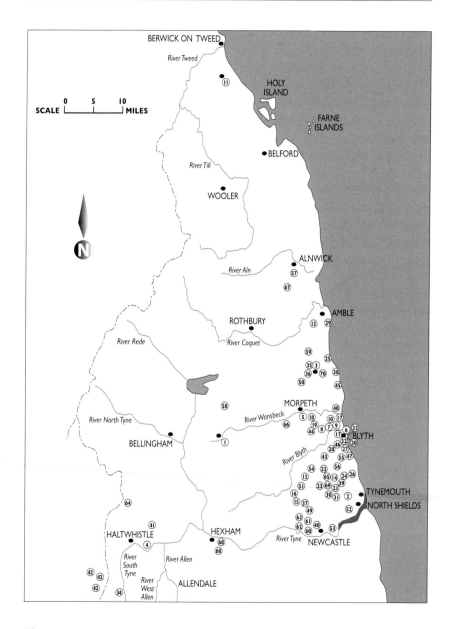

1. Acomb	25. Ellington	49. North Walbottle
2. Algernon	26. Fenwick	50. Pegswood
3. Ashington	27. Gloria	51. Prestwick
4. Bardon Mill	28. Hartford	52. Rake Lane
5. Barmoor	29. Hauxley	53. Rising Sun-Wallsend
6. Bates	30. Havannah	54. Seaton Burn
7. Bedlington A	31. Hazlerigg	55. Seaton Delaval
8. Bedlington D	32. Horton Grange	56. Seghill
9. Bedlington E	33. Isabella-Cowpen	57. Shilbottle
10. Bedlington F	34. Lambley	58. Stagshaw Bank
11. Blackhill	35. Linton	59. Stobswood
12. Brenkley	36. Longhirst	60. Throckley Blucher
13. Broomhill	37. Loughbridge	61. Throckley Coronation
14. Burradon	38. Lynemouth	62. Throckley Isabella
15. Callerton	39. Maude	63. Throckley Maria
16. Callerton Drift	40. Montague	64. Ventners Hall
17. Cambois	41. Morwood	65. Weetslade
18. Choppington A	42. Naworth	66. West Clifton
19. Choppington B	43. Nelson	67. Whittle
20. Crofton Mill	44. Netherton	68. West Wylam
21. Dinnington	45. Newbiggin	69. Williams
22. Dudley	46. New Delaval	70. Woodhorn
23. East Walbottle	47. New Hartley	
24. Eccles	48. North Seaton	

List of the 70 Northumberland collieries operating in 1947.

Chapter 1

Life in a Pit Village

The pit villages in the county were almost all built along the same lines, houses and cottages in terraced rows and as close to the pit as possible. All the men worked at the colliery in some capacity and knew one another. They used the same social, leisure and religious outlets, often subsidised by the coal company. This was the main reason that a special bond existed in the mining communities. Crime was almost non-existent. The ordinary pitman had little of value to steal so doors could be left unlocked night or day. A death in a family was felt by the whole village and no widow or her children lacked support at that time.

After the mines were nationalised in 1947 it was no longer necessary for boys to follow their fathers down the pits as was a condition of employment under early private ownership. Now there was a choice for any lad that left school to take up alternative employment.

Alex Anderson, now a sprightly 93 years, told me about life in Ashington, once the biggest mining village in the world, during the hard times of the pit strikes in the 1920s:

Bedlington Doctor Pit in close proximity to the local pit community.

'Me dad was a miner at Woodhorn Colliery and our home was a small downstairs flat, me sisters slept in one bedroom and me brother and me shared the other with our parents. There was no hot water, washhouse or bathroom. Demand for coal then was sporadic and what with the strikes in 1921 and 1926 our lives were centred round poverty and distress before the days of a Welfare State. But this brought with it a community spirit not known today.

'Food was in short supply and most families relied on the Working Men's Clubs who held soup kitchens for daily meals. I remember well digging on the pit heaps with my dad in a desperate attempt to find fuel for our fire. Me dad was a hard man, he had to work, there was nay stopping off work sick then. Aa've seen him come home at night and he had boils and sores under his arms which me mother had to clean and dress. There were no pit baths at Woodhorn Colliery till the 1930s so mother boiled a big pan on the fire and me dad, brother and me washed in a

A miner's wife filling up the tin bath for her husband. At some of the county's pits it was the 1950s before pithead baths were finally available.

tin bath by the fire. I was the youngest then and had to wait my turn and often fell asleep.

'Like every other pit family in Ashington we had very little but what we had we shared. One day I remember we were all sitting down to eat after work. Mam had made a leek pudding when Doctor McLean called. He said to mam, "By, that smells good, Mrs Anderson, what have you got there noo?" "It's a leek pudding, doctor, would you like some?" "Yes, please," he said, and he sat down at the table with us and she gave him the last in the dish.'

Margaret Nicholson spoke about Lynemouth's long-serving Dr Tom Skene, who told her this story:

'After surgery one day I went down to the newly-opened chemist's shop and found young Bill Tarbit serving a miner who had just moved here from one of the primitive pit villages where there were no modern toilets. Tarbit asked him what was it he needed and the man said he wanted a roond roll of paper for the lavatory. "Just ask for toilet roll then," Tarbit said. The man said, "Alright then, and can Aa have some soap as well?" "Is it toilet soap you want then?" Tarbit said. The man looked at him and replied, "Whey, no man, it's for me hands and face." '

Eileen Brown recalls her time as a young girl at her home in the Hirst district of Ashington during the 1950s:

'I watched for me dad coming from the pit. I could see him from the bottom of our row in Ariel Street making his way down the back lane; him and his marra both as black as crows. They were cutter men who worked on the coal face cutting machines which was a hot dirty job. He often took me with him to Ellington Colliery in the pit tanky when he went for his pay on a Friday. The tanky seats were filthy with the dust from the miners' clothes who travelled back and forward to work for even then many of them didn't use the pit baths, but I felt great 'cos the miners made a fuss of me and I came home with lots of pennies they gave me after they got their pay.

'The pit canteen was next to the baths then, up some

stairs and a dozen old tables and chairs were scattered around the place. I remember the walls were covered with shiny white tiles. My dad and his marras sometimes bought pork pies to eat and on occasions they were so hard they could stot them on the table. Then there would be a good-hearted slanging match with the canteen lasses who gave as good as they got. Most folk only went in for a cup of tea, which my dad described as not just weak but helpless. It was later when the new canteen was built that meals improved and many folk enjoyed a first-class meal there.'

Gladys Aynsley was born and bred in Bebside during the depression years of the 1930s and has lived in the same area for over 70 years:

A typical view of an old colliery village with back-to-back houses and standing off, the netties and ash middens.

'I lived with my mother, father and two brothers in a house at Kitty Brewster Square just behind the pub of that name which fronts on to the main Blyth road.

'The houses didn't belong to the colliery, they were privately owned. We only had two bedrooms. I slept downstairs with me mother and father while me brothers slept upstairs in a room accessed by a ladder. Me father was in the night shift for 16 years and so most of the time I had plenty of room in the bed with just me and mother. The landlord, old Green, was a funny bloke, he thought he could visit the house anytime he liked. One day me mother

---

A hard life for a miner's wife in the 1920s. As well as feeding an often large family on little wages and seeing her husband and sons off to the pit, it was time for washing and come rain or shine that day was always a Monday. Elizabeth Doleman on the left helps her mother to 'poss' clothes in the old tub in the backyard of their house in Amble.

The first pit canteens were basic but clean and well patronised on pay day Friday. Pie and peas washed down with a mug of tea was a miner's favourite.

had chops cooking on the fire when he started to sweep the chimney from the top and the soot was flying all over. She played holy hell and he was careful about what he did after that.

'Me father did work at the Isabella pit for a time and he always said it was the wettest he had ever worked at. Me mother used to scrape the slek off his fustins with a knife before she dadded them against the back yard wall and washed them. Me father worked most of his time at Bedlington 'A' Pit as a rolleywayman. Me mother was a Bedlington woman having lived in Shiney and Telegraph Rows. Grandad Tom was compensation secretary for the Bedlington Doctor pitmen for 33 years and had started work down the pit at 11 years old. Me mother told me he would seek the compen money from Morpeth and carry it home in a poacher's pocket in his coat. Men who were entitled to compen would call at the house and he took them into the back room and doled out exactly the right amount.

'One time I remember well was when me father had

pneumonia and me mother's friend advised using poultices on his chest and back. The doctor, old Fothergill, had frowned on this remedy, tut tutting away saying it was nothing but glycerine and some other concoction. But father got much better and one day mother went down to the surgery and there in the doctor's big cabinet wily old Fothergill had bottles made up of the very stuff he had advised against.

'We had little in the way of worldly goods but were happy with our lot. I attended St Catherine's church and the mother church at Horton and we young ones made our own entertainment. One thing I did miss out on was the Bebside Gala held every year for the mineworkers' bairns. Because me father worked at Bedlington, I was not allowed to join in the local celebrations, I suppose the Bebside miners paid money out of their pay for their own Gala.

'The Co-operative Wholesale Society – better known as

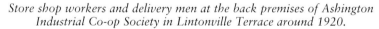

Store shop workers and delivery men at the back premises of Ashington Industrial Co-op Society in Lintonville Terrace around 1920.

the Co-op or the Store – first opened branches in the county in the latter part of the 19th century and by 1925 was well established in most of the pit villages. The Store became a favourite with miners' wives, with most members having a store checkbook and check number. Dividend was paid on purchases every six months.'

Iris Foster remembers Lynemouth Store and the part it played for the mining families and still remembers her mother's check number:

'There were seven Co-op shops in Lynemouth by 1950, all branches of the Ashington Industrial Co-op. Green-grocer's, grocery, butcher's, hardware and drapery were all there. You could buy anything in hardware from nails to carbide to fuel the miners' lamps and the drapery sold all the miners' clothing requirements. A girl called Deirdrie went round the village taking grocery orders and Jack Rowell delivered all manner of goods on a Friday with his horse and cart. When he changed to a motor van some folks played war 'cos they were used to collecting horse muck from the back lanes for garden use. There is still a grocery shop in the village now, the North East Co-op, but all the rest have gone, unable to compete with the big town supermarkets.'

Wine, Women and Wobbly Chicks

Doris Teasdale was born in the model village of Lynemouth in 1925. She never did experience the harsh living conditions of the older colliery villages with their cramped living space and primitive sanitary arrangements with outside netties and middens:

'My dad was a miner and union man at Ellington Colliery and was allocated a house free of rent in Eden Terrace and always told me how lucky I was to have electric light and a bathroom with plenty of hot water. Dad was also given a concessionary load of coal once a fortnight.

'However, the 1926 strike began and I can always remember him telling me that even then there was the odd pitman blacklegging at some of the little pits up north. The depression years of the 1930s followed which meant we had little money. I had a paper round and used an old bike to deliver newspapers to Ellington and Cresswell. I also had to feed our hens that dad had at the allotments morning and night.

'Funny thing happened one day when my mam was making wheat wine. She scraped the leftovers from the winemaking into the hens' feed bucket and mixed it in with their normal crowdy. At night I fed the hens and chicks, then next morning the chicks did not look right and the hens and the cockerel were unsteady on their feet. I brought the chicks over home in a pail, and mam laid out two trays covered with paper and placed them near the open fire. When dad came in from work he said, "What's them chicks doing there?" I told him they were falling all over the place and so were all the rest. "What did you feed them with?" he asked.

"Just the normal stuff," mother said, "and the scraps from the wheat wine."

"Nee wonder they are falling about then is it, 'cos you knaa what's the matter with them – they're drunk as monkeys!" And we all had a good laugh at that.

'There was one time during the depression years of the 1930s when the men were on short time at the pit, the Lister family held a raffle at the Lynemouth Social Club for joints of meat from the pig they had just killed. Well, my dad won a prize which they said was a yard of pork. Me mam said that's great, we'll have plenty of meat to last us for ages. One of the Lister lasses came to the door with a parcel of meat. Dad opened it expecting a big joint of meat but instead it was a string of sausages.

Later colliery houses were built on a grander scale than the back-to-back type in the old villages. Lynemouth houses built by Ashington Coal Company from 1922 contained from two to five bedroom accommodation according to the size of the family. An internal bathroom with plenty of hot water on tap and a large garden to the front made life much cosier for the mining families.

"That's not a yard of pork," he said, "Aa'm going down to see Lister."

'And he did, saying to him: "Aa must have got the wrong prize 'cos that's not a yard of pork I got, it's a string of sausages."

"Whey, aye it is, man," Lister said, and measured the string of sausages. "There you are you see – exactly one yard long – there's your yard of pork sausages."

'Me dad was gobsmacked and what could he say 'cos it was pork, he just came away and had a good chuckle.'

Flying Pigs and a Dancing Horse

Alan Reed never forgot his younger days and the characters in his home village:

'During the dig for victory campaign during World War Two, many Lynemouth families worked allotments producing their own food for the table, growing fruit and veg and keeping hens for eggs as well as fattening pigs for a meat supply. One Saturday my pal and me went along to the allotments to see his dad's pigs being killed. The butcher, Bob Lunn, was going to use the new humane killer gun which replaced the old method of a spike driven into the pig's head.

'Quite a number of locals turned out for the event and boy, they were not to be disappointed. The pig sty lay at the bottom of a steep hill away from the road but they decided to kill them on the road where a horse and cart was waiting to take the two pigs away for preparation. A group of men put a rope round the first grunter and, pushing and heaving, eventually manhandled the beast up the hill and onto the road, its rear end facing back the way it had come from. Then the butcher placed the gun on the pig's head and fired, which sent the pig flying backwards and ending up stone dead outside the pig sty. This meant it had to be dragged back up the hill to the road with a great struggle.

'The men had not learned a lesson from the first mistake and followed the same procedure with the second pig. Once again after it was shot it also flew back down the slope to the sty where it had to be dragged back to the road. An argument raged as to whether the two pigs would be too heavy for the horse to carry but one man shouted: "Whey, not likely, man, put them both on."

'They did just that, but when the horse moved away the cart hit a rut and the pigs rolled back and the poor horse was left dancing in the shafts in mid air frantically trying to get its feet back on the ground. Everybody rushed to move the pigs forward and get the horse back on its feet and stop the shafts from breaking. I think a lot of folk learned a lot that day.'

Silver and Brass

Most colliery villages had their own band, originally named Silver Prize Band or Silver Model Band, which evolved in the middle of the 20th century into the Brass Band we know today. Many pits created a Miners' Union Lodge Banner which was paraded at the head of any miners' gathering, whether it was a show of support in time of strike action, a children's gala or any village festive occasion. There is nothing that stirs the blood of any ex-pitmen more than to see their own pit banner being borne aloft and followed by their colliery band with a lodge of miners marching behind. This is a legacy that was handed down from the bitter struggles between the unions and coal owners of the 19th century.

Ashington man, Jim Locker, is now 85 years old and has been a bandsman since he was a ten-year-old lad, when he joined his father and two brothers in the Public Band in 1933:

An early 20th-century photograph shows that like all other bands at that time there was no money for uniforms and many of the instruments were purchased by the miners themselves on the 'never-never' scheme. Later the unions and the coal companies added much-needed support.

NETHERTON COLLIERY SILVER MODEL BAND.

T. COLE. PHOTOGRAPHER.

'Every night when me dad came home from the pit he would make us practice. I played the baritone them days and me dad played a rare left-handed cornet specially made for him as he had lost a limb in the Great War. The Ashington Group of Collieries wouldn't sponsor us and so we had to beg, borrow or steal to keep the band going, you know, shaking collection boxes, knocking on doors and things like that.

'Our uniforms were ex-policemen's with gold braid attached. Things got better when North Seaton Colliery officials asked us to become their band in 1935 and gave us financial backing. Conductor Sammy Bond could get a man with no arms to play the trombone, he was that good. By 1937 our young band won the Northumberland Miners' Picnic and again in 1938 before doing the hat trick in 1939. I was playing the trombone by then but my arms weren't long enough so they made me a wire attachment to reach the seventh shift.

'We lost one member to the Second World War: Jackie Turnbull who played soprano, he got back from Dunkirk but died in hospital. We played at Jackie's funeral, that was hard for us and the biggest funeral I have played at.

'Success continued after the war with wins in 1946 and 1947 when we played in the British Championship at the Albert Hall. When North Seaton Colliery closed in 1961 we returned to Ashington as the Colliery Band. One of our highlights was recording the Hovis advert on TV for which we got £600 and that was a lot of money in the 1970s.

'Ashington Colliery closed in 1988 and we came under the umbrella of Wansbeck Council who provided financial backing. I've always played in the same band and guested for others when needed. Sadly, there are only a handful of bands now and all those are sponsored by other sources than the pits, which have all closed in the county.'

Doris Little also looked back:

> 'My dad loved his music and played the violin while me two brothers were both bandsmen with the Ellington Colliery Band, one played the cornet and the other the tenor horn. When they were practising in the living room the noise was deafening and I went outside to play.
>
> 'I married Norman Teasdale from Lynemouth who worked at the pit and played in Ellington Band. He had two accidents at the coalface and had been happed up twice. After the fire at Lynemouth in 1966 I persuaded him to leave the pit and we moved away to the Midlands. As soon as we arrived, folk came knocking at our door asking Norman to join their band which he did – it was his life – and he played for the Royal Doulton and other famous bands. Every time we went back up North he would never miss a practice with the Ellington Band. We moved back to

North Seaton Colliery Band in the late 1940s outside their headquarters, pictured with a wealth of silverware. Conductor Sammy Bond centre front and directly behind, Jim Locker.

Newbiggin in 1972 and he played with their band and guested for others when they were short.'

Big Leeks

Leek growing, and pigeon and whippet racing were widespread throughout the county pit villages – a legacy from Victorian days when coal owners would indulge in their sports of hunting, shooting and fishing, then so too would the ordinary pit lads have their own brand of leisure activity.

Leeks, of Middle Eastern origin, are one of the oldest cultivars in existence, being mentioned in the Bible, and were introduced to this part of the country by the Romans. The Romans also mined outcrop coal, so in effect it was they who were the first leek growers and miners in the county almost 2,000 years ago. They would no doubt have been amazed to see the giant show-leeks grown today. Leek growing in the pit villages began in the late 18th century, at first purely to put food on the table, but by the middle of the 19th century it had developed into a competitive hobby for many pitmen after spending long hours in the bowels of the earth.

Iris Foster of Lynemouth looks back:

'At the early leek shows at the Miners' Institute, the Hotel and the Social Club prizes were always goods supplied by local shops such as the Co-op and other big stores from Ashington and Newcastle. Things like fridges, coffee tables, and standing lamps. Sometimes folk would swap prizes if they didn't like what they had won and then the big stores got sick of the hassle of it all and from the 1970s it all changed to money prizes.

'Lots of the pitmen showed leeks, my mam's two brothers and their dad were all leek men. The women never disturbed the men while they dug them out. The leeks had to conform to show standard and often this meant stripping the leeks down and leek flags were scattered all over the back garden and tempers could get a

29

Joss Hanson and uncle John Graham were consistent leek growers and winners at Lynemouth in the 1970s and 1980s and are pictured here in the back yard of the house with their stands of leeks.

little frayed at that time trying to get a stand of three leeks for the show.

'After my Billy had dug his leeks out and cleaned them at show time, he brought them into the back kitchen and I would buff them up with a cloth soaked in milk to make the barrels shine even whiter. They would then stand overnight on the table covered with a damp cloth to stop them drying out. The night before the show was when the banter started between the miners. Meeting up at the village pub, none of the growers would admit to having good leeks, saying it had been a bad year and they had no chance at all of winning. Secretly every one of them had high hopes of winning the coveted trophy and the top prize that went with it.

'Next morning the leeks would be checked for splitting and measured again to make sure they were not above the required five and a half inches long to the fast button, which was what was allowed in Northumberland. Then the men would carry them down to the show ready for judging.'

Bob Snaith, a joiner who worked at Ellington, had this to say about leeks:

'The breed of leek went a long way to bringing success in the shows. Me and me marra one year in the 1970s got half a dozen leeks each off the top grower at that time, Joey Jones of Stanhope. I won my local show at the Northern Club in Ashington with a stand of three leeks out of the six grown and so did my marra at his show, you can't get much better than that 'cos some of the lads were growing 30 and 40 leeks in their trenches.'

Bob Bell from North Seaton Colliery is now in his seventies and has grown leeks for as long as he can remember. He now judges the World Championship Leek and Onion Show each year at Ashington:

'Me and me brothers after finishing school each night had to dig or plant wor own section of garden before we were allowed oot of the hoose. Me father was strict aboot that, and that's how me and lots of me marras in the village started with leeks.

'It was handed doon from father to son. Every grower

A world record and into the Guinness Book of Records *for Bob Bell with this monster heaviest leek grown in 1989, displayed here in the front of his workplace at Alcan's aluminium smelter, Lynemouth.*

had his own potions or secret recipe for feeding the leeks. Me father would send us oot into the fields to collect sheep's droppings, put them in a muslin bag and then scaad them with boiling water. They were then dropped into a water barrel and stank to high heaven. The foul smelling liquid was used to feed the leeks during the growing season. Some folks had the notion that peeing on the leek bed would help them grow, but it was more likely they were caught short on their way back from the pub. The methods and strains are different noo, one time they were grown outside but the day it's all polytunnels and regulated watering and the leeks grown to huge sizes.'

Cocks and Hens

Pigeons have been part of a miner's life since the 19th century. Kept in the first instance in back yard crees, and later on allotment sites, they provided a fresh source of food in desperate times during the many pit strikes and lockouts. Sad but true, that it came to that in the bad old days of private pit ownership. Pigeon racing really came to the fore in the early 20th century and provided an outlet for the miner after working deep in the bowels of the earth. An affordable sport then, with cheap corn available and most pitmen could knock up a cree with old doors and wood in next to no time. Today it is different, a high tech outlook on feeding, training, methods and the lucrative prize money available attracts folk from all walks of life to the sport.

Jimmy Swan of Ashington tells it as it was, and is:

'I was only a laddy about eight or nine years old when I

Jimmy Swan holds a pigeon from an auction he held to raise funds for a new Community Centre in North Seaton; on the left is local Councillor Jim Weallans and to the right MP Will Owen.

timed in a pigeon from Arras for the Bell Brothers of North Seaton Colliery. The brothers had gone home for a bite of breakfast and what a thrill it was for me to handle that bird. I reckon that's what got me hooked on racing pigeons. This bird had just flown some 500 miles across the English Channel and home, and that to a young lad was an amazing feat.

'At first the members had no clocks of their own and used young lads to run to a central clock held by George Davison at First Single Row. Every fancier was allowed so much time according to where their pigeon loft was situated from the clockholder. The idea was when a pigeon landed, the rubber ring with its own special number on was taken off its leg, placed in a thimble and the boy runner ran like the clappers to get the pigeon's time recorded.'

Bob Harmer was the clocksetter and what he didn't know about them wasn't worth knowing. On rare occasions if any fancier cheated by manipulating his time, Bob knew how it had been done and the man involved was shown no mercy. He said:

'I started racing as a teenager after World War Two with Ken Gray and on my own in 1949. Over 60 years later I am still at it although on a much smaller scale. We had some great fanciers then with the likes of Taylor and Hay and Richardson Bros who won an Olympiad gold medal with their famous pigeon *Iron Man*.

'We raced in the Wansbeck Federation and to get them to the race point a lad called Norman Glass would take the baskets of pigeons to North Seaton station on the back of his horse and cart. We couldn't afford to send many pigeons, not like today where some of the big flyers send as many as 40. That was about the number sent by the whole club in the old days.

'I have had some good birds through the years and one

in particular was a mealy hen who won me over £800 in 1971 and from the proceeds I bought a new car. I loved to get involved with the admin side and over the years have been Wansbeck Fed Secretary, Combine Delegate and Vice Chairman.'

Kit Miller remembers one Thursday when he was manager at Whittle Colliery during the 1970s:

'It was not long after I had started work at Whittle that a man came to see me on a Thursday asking for a rest day on the Friday. I asked him why and he said he needed to basket his pigeons for the weekend race. "Where are you flying from?" I said. "Beauvais in France," he replied. He did not know I was a pigeon fancier and I knew the pigeons were always ringed and basketed on a Thursday for the Channel races. "That's funny," I said, "'cos my birds went away this morning." His face went red and he just shook his head, mumbled and crept out of the office and I had a good laugh about that.'

Sport was always on the agenda for the miners who spent a shift in the bowels of the earth breathing in fumes and black coal dust and then came to bank lusting for fresh air to fill their lungs. Football, cricket, hockey, bowls, rugby, tennis, boxing and gymnastics were all available. The coal companies in the 1920s and then the National Coal Board (NCB) from 1947 looked after the social welfare side of mining communities, building institutes, welfare halls and providing playing fields at venues around the county.

Each pit had at least one football team and often many more from different departments of the complex. Football was the favoured sport and so popular the crack was that if any team needed a player for a match they just had to shout down the nearest pit shaft and out came a miner to fill the team sheet: Jack and Bobby Charlton of 1966 World Cup fame, whose father worked at Linton Colliery, played in the local Ashington Welfare League; and there was Bob Farrington of Lynemouth and his son John who played for Leicester City; John Angus from Amble, a stalwart for Burnley; or Mal Musgrove, a Lynemouth Colliery clerk who played with West Ham in the era of Bobby Moore and was spotted playing in the Welfare League. In the latter years of the 1970s and 1980s, Newcastle-born Brian Little hit the headlines with Aston Villa before heading off into management and becoming a TV soccer pundit, as well as managing Wrexham. Cec Irwin was an Ashington lad who graced the Roker Park turf with Sunderland for many years from the 1960s. But probably the best known and respected was the Ashington man and hero of Tyneside, Jackie Milburn. A pit fitter at Ashington and Woodhorn collieries in the 1940s, he went on to play for Newcastle United and England with a goalscoring record for the Black and Whites that was only broken by the latterday Tyneside legend, Alan Shearer.

Jackie Milburn never forgot his mining background and lived and died in his home town of Ashington. Jack is on the right of the picture having a crack with some of the lads after their shift shortly before the pit closed in 1988.

Joe Grieve pictured centre with the ball was Secretary of the Miners Hirst Welfare in the 1950s and 1960s and also ran an Ashington Junior Team that took all trophies in the East Northumberland Junior League in the 1959-60 season.

I remember one time when I was at school at Lynemouth in 1954 and Newcastle were having a successful Cup run. My pal Jimmy Graham and me asked our teacher Mr Reekie if we could have the Wednesday afternoon off to watch them play Huddersfield Town at St James' Park. He said no, but we took the day off anyway.

The next day at school we thought we had got away with it, but after dinner the Head, Mr Graham called us up to his office and gave us ten of the best with his cane. It hurt like hell but was well worth it as the 'Toon' beat Huddersfield and went on to win the FA Cup that season.

Chapter 2

Gannin ti thi pit

My old friend Tom Nesbit has lived in the pit village of Ashington all his life and has written many poems about pit people and how they lived. Here is an extract from his poem *Miner*, which captures the atmosphere as he leaves home to go to work in the dead of night.

Aa remember me pit boots echoing
Down a one lamp street.
A winter's night,
Gannin ti' pit
Foreshift,
Me bait tin rattling thin,
Only the warm tea bottle
Wi' any sense of discipline.

A single Woodbine passed around
Before the cage
raged
against the darkness
of the hole
that led us all the way to hell.

Miners worked an eight-hour, three-shift system from 1910 and Northumberland was one of the last places in the country to conform to this practice. Foreshift meant the starting time could be anywhere from midnight through to 3 am; Backshift from 7 am till 10 am; and Nightshift from 2 pm until 5 pm. Foreshift and Backshift were considered the main coal production shifts. Nightshift was for repair and maintenance and named the 'Owld Man's Shift' simply because young lads disliked it as it spoilt their social life.

Backshift meant work, social and rest time could all be had in equal amounts but Foreshift was named the 'Graveyard Shift', going out at the dead of night when other folks were tucked up in bed. There were those who kept this shift as it suited their needs, starting on a Sunday night and ending on Friday morning, but most of the lads just could not sleep through the day and could not be bothered to have any social life in the early evening. It was easy to have a couple of hours' rest or doze on the settee late evening but many were so tired they 'slept the calla', meaning they didn't hear the alarm clock and so missed their shift which meant losing money.

The bait or snack was put up by the miner's mother or wife and often consisted of jam or dripping sandwiches hurriedly washed down with a bottle of water, halfway through the shift.

Before me dad left for the pit he had a ritual he practised which saw him recite, 'Bait, bottle, baccy, matches, tabs', and he would pat his pockets to make sure they were all there. He told me that the matches and tabs were hidden on bank after he had a last drag at the shaft top and so were there when he came back out the pit when he was gasping for a smoke. The baccy he took with him underground. There were two reasons for this: one, like lots of other pitmen he would snap off a twist of baccy and chew it to take away the craving of wanting a smoke. Secondly, a 'chaw' of baccy kept the mouth moist and it was common to see the older pitmen spitting the brown juice out the side of their mouth.

To clear the head underground, snuff taking was a common practice. The regular users carried it in a little tin box from where

A typical miner of the 1920s and 1930s ready for work with his soft cap, jacket, waistcoat, short trouser fustins, thick blue stockings and boots, and holding his oil-filled Midgie lamp.

they took a pinch and spread it along the back of their hand before sniffing it up the nose in one go. I tried this only once when old Ascot persuaded me it was good for the head: 'Clears the airways,' he said, 'gets rid of the dust.' Well it did more than that, it nearly blew my head off, brought tears to my eyes and I couldn't stop sneezing.

A well-documented story is where a newly-wed miner complains to his wife about having only one sandwich in his bait tin. The next shift he gets two and still complains so she puts up three and he still complains every day even when he has a tin full. She decides to teach him a lesson and cuts a whole loaf of bread in two and sticks it together in his bait tin. When he comes off shift she said, 'Did yi enjoy ya bait the day then?' He replied, 'Yes, pet, Aa did, but Aa see ya doon ti one slice again!'

Geoff Murray remembered an old character who worked at Ashington Colliery:

'When he was woken by the alarm clock about midnight to go to work in Foreshift, he was known to say to it, "Aye then, you've woken is up again eh, well Aa'll tell you what

*A team of drawers about to descend, waiting at the shaft top at
Woodhorn Colliery in the 1930s.*

– if you're that keen you can go to bloody work yasel."
Then he would turn over and go back to sleep!'

A bloke called Carl lived in Lynemouth and worked at Ellington Colliery which meant a mile walk up a dark road. He would meet up with his marras at the bus stop. One Foreshift he was early and waited inside the bus shelter and fell asleep on a seat in a corner. His marras, not knowing he was there, went to work and two hours later Carl woke shivering with cold. He couldn't go to work as the last cage would have been long gone. He went home and told his missus he had been sent home with flu and she banged the aspirins and hot drinks into him. Next morning his wife was on her way to the doctor's to make him an appointment when she met one of his marras coming home from the pit. He said: 'What

41

happened ti Carl last night? He never turned up for work, bet you were sleeping on his shirt tail eh, Betty!'

She went home and gave Carl a good tongue lashing, demanding to know where he had been and was he cavorting with another woman. Finally he had to admit he fell asleep and missed his shift, but it would have been better if he had come clean in the first place.

Alan Young rolled back the years when he told me about an incident from the 1950s:

'Old George was walking down the pit road from his house in Lynemouth to start his shift at midnight. It was a cold winter's night and he had his head bowed against the wind. Suddenly he crashed into a figure that appeared in front of him carrying an axe on his shoulder. He almost collapsed with shock until the person steadied him. It was only Jimmy, a face drawer, coming home from Nightshift who never left his axe in the pit. George gasped, "By, yi give me a turn there lad, thought yi were the grim reaper! There's na way Aa can gan ti work noo," and he promptly turned around and headed back home with Jimmy.'

Chapter 3

Doon Under

Stepping into a cage for the first time to go underground was often a terrifying experience for young would-be pitmen at 14 and 15 years old. Prior to the Second World War there was little or no training for the job in hand. This situation changed dramatically when training facilities were set up both on the surface and underground by the NCB in the 1940s. Gone were the weeks of training for life deep in the bowels of the earth, now it was time for the real thing. Crammed into a cage full of men and boys carrying lamps, explosive boxes, bait bags and tools, they had to endure the taunts, earthy humour and initiation tests of the older and more experienced pitmen.

Somehow the winderman who operated the controls would get to know that a rookie young lad was on board and would set the cage away at speed, then brake so that the cage pulled up sharply and bounced up and down in the shaft on the end of the winding rope. It had the instant effect of making the heart race and caused a sinking feeling in the pit of the stomach. This was the signal for the older men to poke fun at the by then pale, wide-eyed young 'uns by saying: 'Aye lad, that's right, this was the same spot last week when the cage got stuck in the shaft, aboot 300 feet from the pit bottom and we aal had to jump oot onto the shaft ladder and climb doon. Pity though aboot me marra, he fell into the sump and he's niver been seen since!'

Ron Simpson recalls his first day as a Bevin Boy at Choppington 'A' Pit:

> 'When the winderman set the cage away it went steady, then suddenly dropped down the shaft like a stone and almost lifted our legs off the deck and I lost no time getting out at the bottom. I had been given the "greenhorns' ride" which as I learned was given to every new starter.'

There are lots of mining songs and chants, some of which date back to the very early years of mining. The Northumberland and Durham coalfield is the oldest in Britain and possibly the world and many of the old songs stem from this area, fuelled by the harsh conditions and bitter feuds with the pit maisters. Here is a relevant one that is well known throughout the mining communities:

> *Don't go down the mine Daddy*
> *Dreams very often come true*
> *'Cos Daddy you know it would break my heart*
> *If anything happened to you.*
>
> *So tell my dreams to your friends, Dad*
> *Go tell my dreams to them all*
> *For sure as the stars that shine, Dad*
> *Something is bound to befall.*
>
> *So Daddy don't go down the mine*
> *Oh, Daddy don't go down the mine.*

My Uncle Bob was a Deputy at Lynemouth and like some of the older pitmen, was very religious. They would never be heard to swear or curse which was amazing when things could go drastically wrong in a hostile environment. If things were going badly they would use words like 'Yi blowdy', 'Blummin' heck' or refer to someone who upset them as 'that beggar'. Bob was a Primitive Methodist preacher on the Ashington Circuit and never lost that faith. One particular incident in the 1960s was when he was in charge of a district which had seen many accidents, and where roof falls were a common occurrence. Before leaving to work on the coalface he gathered his men at the Deputies' kist and held a little prayer meeting. Not a sound could be heard, every man listened to his words and gave him the respect he had earned.

Bob Bell started work at North Seaton pit way back in 1950 and remembers it well:

'Me dad, brothers and uncles all worked at the pit. It had a homely feeling 'cos most of the men in the colliery raas worked there and so everybody knew each other. It was hard graft but I was happy there. The main shaft was fine but the second upcast was a primitive affair. The pit was hotching with rats and yi had to carry your bait in a tin to

North Seaton Colliery, owned by the Cowpen Coal Company, opened in 1859 and closed in 1961.

stop them from eating it. Aa kept the Foreshift for seven years, it suited me fine gan ti work at midnight and finishing for the week on a Friday morning.'

Geoff Murray worked at Ashington and remembers his time when training on the coal face as an 18-year-old with experienced pitmen:

'The Training Unit was a new face just starting away and after a few weeks the goaf or worked-out area stretched a long way back. This meant it was waiting for its break, a collapse of the roof to settle things down. Thankfully for me, I was in Backshift when it happened one Nightshift. One of my marras who was there said it happened all of a sudden and the goaf came down with a sound like thunder. It didn't just stop there, it skittled out planks and props on the face, and men and lads were scrambling to get off and into the roadways. He told me his trainer old Buck who could barely walk passed him like a shot out of a gun. He said it was the most frightening thing that he had ever experienced.'

Alex Anderson recalls:

'I was 14 years old in 1928 and left school in Ashington on the Friday and went straight up with me dad and brother to sign on for work at Woodhorn Colliery. My mining life started on the Monday at 2 am when I went underground with my dad and brother. I was put to work at the shaft bottom where it was freezing cold, in fact in the winter braziers were lit at the shaft top to stop surplus water from freezing. Injuries were common them days and my brother damaged his fingers on an arc-wall cutting machine. This could have been a serious blow to the family income so I agreed to take his place to put food on our table. I had no training on this lethal machine but survived for two weeks

A typical bottom of the pit shaft scene showing the onsetter with the miners in the cage ready to come to bank.

till he returned. I was 21 when I was given my own coal cutter and a young assistant, then switched to long wall where the seams were only from 16 inches to 3 foot high.'

Stan Elliott worked at Ellington Colliery for 38 years and said:

'I followed me dad into the pits; he had worked down the pits in the Whickham area. I did all the grades of pitwork from filling tubs to machine work. There was a great camaraderie that existed between men underground working in uncertain conditions where danger was ever present. Besides the obvious dangers of gas, water, and roof falls, there were so many other things that could happen.

'I remember one incident in 1960, doing a routine job when an event occurred. Me marra and me were on stonework up the North West Diamond District. We needed to split a prop to support an arch girder over the Barrier Roadway. I held an axe on the prop and me marra hammered the top with his axe. Something struck me in the chest and I knew then what it was like to be shot. I went down on my knees like a shot cock not knowing what had happened. All I could see was a little hole through my vest and blood pouring from a wound. I had no idea then that the steel shard had lodged itself well into my chest. The Deputy didn't think it was too serious, saying "Yi'll be aalright, man", bandaged me and sent me outbye to walk all the way to the shaft, a distance of about two miles. The nurse in the Medical Centre sent for an ambulance and at the hospital they took a big sliver of steel

A last breath of fresh air and a canny crack before their shift underground for these North Seaton men in the 1950s pictured outside the pit baths: Jimmy Lyall, John Scott, Tom Soulsby, Ron Lillico, Jim Swan, Willie Alexander and Jimmy Gray.

from the wound. I was a very lucky man that day, the doctor told me that the steel had lodged just a fraction away from my heart.'

Bob Charlton started work in the pits in the 1960s straight from school:

'Me dad worked at Woodhorn Colliery but I went to Ellington cos all me marras were there. The camaraderie was second to none and everybody looked after one another. When I started to work with ponies down the East Arterial flonking girders in to the miner flats came an incident that showed the dangers to be faced. I had a pony who was a bit skittish and he took off one day with girders attached and hit a big swally of sleck and water. He just disappeared under it all and then his head appeared and we had to feel under the sleck to lowse his traces off the girders. When I got him out, him and me were caked in the stuff.

'I worked a lot in water sometimes up to my waist and the owld pitmen said, "Yi'll suffer for that lad, later in life" and it came true as I have a bad time with arthritis now. Later I worked on the JCMS and eventually took on a beast of a machine called the Heliminer from 1982 which cut coal like it was going out of fashion. Our set broke many production records and we made good money but worked bloody hard for it, earning bonus payments which could take us up to £400 a week. It wasn't all work and no play, we had our moments when the belts were standing and played tricks on one another and things like that.

'We had a good set of lads who would help each other no matter what. If one was feeling off colour his work would be done and no questions asked. Can you imagine that happening today?

'You had to have your wits about you and once when me marra Ian was on the machine handles I noticed the roof

Deputy Jack Boaden writes out his shift report at his kist under cover of brattice cloth in what he describes as a very wet district of Ellington.

flaking a bit and told him to stop the machine and come out of there. Shortly after, the whole roof came down and closed the place for some time. We were working all kinds of different shift times then, even starting at 4 am in the morning which is no good for any man. We had some good bosses and some bad ones, the best were the likes of those who had come through the ranks like Tom Smith and Jack Tubby, they knew pitwork and the men and what they could do. You could sit down in their office and negotiate with them, put forward your point of view and we would try different methods of working and things like that.

'I can remember lots of incidents when lads were hurt or killed and looking back I sometimes think your life is mapped out in some way. One lad was killed by a fall of stone because he changed shifts with his marra and another when he stayed back for some overtime. A lad called Robinson had his back broken and was in a wheelchair for the rest of his short life. There was no laying the pit idle as a token of respect like in the old days, you just got on with things.'

Mike Kirkup started his working life down Ashington Colliery as a young lad and remembers it well:

'I was never cut out to be a pitman, not like my dad Jack or his brothers Larry, Sid and Denis; and certainly not like my mother's six brothers, the Talbots, who mostly worked underground at Woodhorn Colliery. I lasted for five years underground and reckon that was a long enough sentence to serve.

'That was the thing about living in Ashington or any of the other surrounding towns and villages in the 1950s: a lad's life was all mapped out for him while he was still in short pants. The decision was not his to make; there were few options.'

Chapter 4

Pit Humour

When working in the bowels of the earth with only a cap lamp for illumination and often in wet and extremely dangerous conditions, having a sense of humour was one way of making the job that little bit more bearable. Every coalfield has its own brand of pit humour that is often as raw and earthy as the coal itself.

This classic tale often quoted is where two lads working through the night in Foreshift after having their bait in a siding, doze off and are then confronted by the pit manager who has been called out to a problem. Shining his spotlight into one of the lads' eyes he says: 'Hey lad, wake up there, do you not know who I am?' The lad opened one eye and turned to his marra, shaking him and saying: 'How, Geordie, yi had better wake up noo 'cos there's a bloke here who's lost his memory.'

Norman Mathews came to Tyneside as a Bevin Boy (see chapter 6) which changed his life for ever, and looks back over the years:

> 'Pit folk don't take long to cotton on and give someone a nickname if they think they deserve it. I got the nickname "The Undertaker", not because I took on any job available but for the fact that every pit I worked at closed down.'

Alan Reed looked back to his time spent as a storekeeper on bank in some of the county's pits during the 1960s:

'I worked with an older man in the storehouse at Ellington called Harry and what a character he was. Near the end of the shift, Harry would chop nice dry sticks for all the lads and bundle them up as kindling for the home fire. We had a new gaffer who was a bit feisty and could make your life hell at times so Harry decided it was time to teach him a lesson. The Gaffer caught Harry chopping sticks one day and demanded that he should get some or he would stop everyone else taking them home.

'The next morning the Gaffer was late coming into work and said to Harry: "It took me two hours trying to light the fire this morning – them sticks wouldn't catch hold at all. What sort of sticks were they?" Harry straight-faced replied: "Whey, yi got the same as the other lads and they never have nee bother lighting the fire." The lads behind the partition were doubled up with laughter 'cos they knew that he had given the Gaffer the sticks from a wood prop that was still green and wet and full of resin.

'He would play tricks on the coal fillers when they came to the stores with a note for a new shovel which were hung on racks inside the store and he would give the filler a shovel to handle. "That's too heavy, Harry. Nee good for me. Hoo can Aa fill coals with that? Let's try another one." Harry would stomp away and bring another shovel.

"Nee good this one either, Harry. Doesn't have the right balance man," the filler said, swinging the shovel as if he were casting coals. Away Harry went to fetch a different shovel. Coming back to the store's hatch he would swing a shovel with one hand saying: "Look at this one, man. Light as a feather. Balances like a set of scales. Best shovel Aa've ever handled."

'The filler would take the shovel, balance it in his hands and smile. "By, you're right, Harry. Yi knaa a good shovel

"*How is it that you call yourself a distant relation of your brother, Mike?*"
"*Well, ye see I was the first child and Micky was the seventeenth.*"

alright, it's just the ticket, Aa'll have this one." Little did the filler know Harry had given him back the shovel that he had refused in the first place.'

Bob Bell talked about a man he worked with at North Seaton:

'We used ti sweat a lot doon the pit and one bloke Aa worked wi on stonework never took his claas yem ti be weshed. He was hummin', specially his socks, when he tuk them off thi stood up stiff like thi wor frozen. Wi played tricks on him and sprayed him wi scent and he would gan aboot sayin', "By, what's that smell, it stinks?" Then the deputy came in and went sniffin' the air, sayin', "What's that smell?" So wi said, "Oh, it's Herby, man, he's just had his socks yem ti be weshed." '

Kit Miller a home-grown manager brought up in the pit village of Linton, did all the grades of pitwork before becoming a Fore Overman, Under Manager and then Manager at Longhirst and Whittle collieries:

'One day at Longhirst one of my men came to see me on a Thursday, I reckon it was around 1960. This man who had just transferred from another pit came to my office and

Longhirst Drift in 1960: it produced 4,000,000 tons per annum during the 1960s with a workforce of 800 men.

asked to have a rest day for the Friday. I asked him the reason and he said his wife was going to hospital to have an urgent operation. "Certainly," I said, "no problem at all." The next day one of his marras from the old pit they had worked at asked his whereabouts. "His wife's away to hospital for an urgent operation," I said. He started to laugh and said, "Whey man, his wife must have some sort of record 'cos that's aboot 20 urgent operations this year up ti noo."

'Anything to get a Friday off. I remember a lad called Tommy Nesbitt who came from West Sleekburn pit and was a good singer and ended up singing worldwide as part of a famous duo, Millican and Nesbitt. When work was stopped on the coal face where we had the new telephone speaker system installed some of the lads would urge Tommy on to sing and he would burst into song and everybody would listen to him singing at various points along the face. He wouldn't work on Friday nights so I called him into the office and asked him why. "Whey man, Aa can get more money singing in the Club Go-As-You Please on a Friday night than Aa can coming to work at the pit." So I said to him, "Look Tommy, I have a pit to run, you have a choice, you are either a pitman or a singer."

'Later he did leave the pit and at a safety dinner I attended one night he was the star turn. Before he sang he said, "Ladies and gentlemen, there is a man in the audience tonight who once said to me, you have the choice to be either a pitman or a singer and I want to thank him for giving me that choice."

'One bloke called Baggy came from Barmoor, a little pit up past Morpeth. He was a cutterman and a good one but was very temperamental and sometimes if his place of work was not ready he refused to go to another district to cut coal. "You have to go," I said, "your face is not ready." "No, Aa'm not going in there, Aa don't like it." No amount of coaxing would make him see sense and so I had to tell him to put his coat on and go home. This happened a few times and eventually if his place of work wasn't ready he would say to me, "Put me coat on then? Eh, boss?" and off home he would go on his motorbike. He was probably the only man I know that worked at two pits. I learned he was working weekends coal cutting at Barmoor and yet was on my books working at Longhirst through the week.

'In 1965 an NCB film crew came to the pit to make a film called the *Longhirst Story* about us being the first colliery in Britain to produce 1,000 tons of coal a shift off a single coal face. The film crew were living in a Morpeth hotel and brought loads of exotic sandwiches and other goodies for their bait every day. My lads would say to them: "Whey, what have you got for your bait the day then? It looks better than me jam sandwiches." Then they would plague the visitors mercilessly until they shared out the hotel picnic with them.'

Jack Tubby, Ellington Colliery Manager, smiled when he recalled the day before the pit stood down for the Christmas holidays in 1982:

'I decided to go underground that day and as I travelled the shaft roadway to access the man-riding set, I could hear some of the men shouting, "Here comes the manager. Look oot." And then a cry, "It's too late!" as I rounded the corner and came face to face with a small pony dressed as a reindeer, antlers and all. Surprised yes, but I had a good laugh. The pony handler was a relieved man. What could I say anyway? The man-riding train moved off on time and I reckon the pony-cum-reindeer enjoyed his moment of stardom.

'When I arrived inbye I visited a film production unit and came upon a sight I had not expected to see. It was bait time for the early shift and to say it was Christmas was an understatement. The men had erected a table using an old ventilation door supported on wood chocks and decorated it with coloured paper tablecloths. It was laden with food that would have done justice to any Christmas banquet. I was manager but no killjoy, the pit had been breaking production records for some time so I just reminded the men not to take too long at their celebrations and continued on my way, after all it was Christmas, wasn't it!'

Geoff Murray told me of one day down Ashington Colliery in Backshift when old Jack Swanson and three other men were sent inbye to clean up a spillage at a loader end. They saw a spotlight gradually coming inbye towards them and knew it was an official, maybe even the manager. Jack's three marras grabbed the three shovels available and began casting coals onto the belt and into tubs. Jack was left without a shovel and the manager noticed this and said: 'Jack, can you not find yorsel a shovel? It's not like you to be standing doing nothing.' As quick as a flash old Jack replied, 'Aa'll tell yi this, Mister Smeaton, when yi gan away from here Aa'll get thi bloody lot.'

My younger brother Owen Taylor worked at Ellington for 36 years doing all the grades of pitwork before becoming a Deputy and said:

'Pitwork was hard, there was danger everywhere, gas, water, dust, roof falls, runaway transport and much more. No one knew what might happen, and if we hadn't made the best of humorous occasions it would have been a tense existence.

'One time especially was when our set of men were coming outbye riding the belt and a bloke called Bob lost his false teeth among the coals. He stopped the belt and we all searched for ages, but never found his gnashers. There was war on, phones were ringing, the gaffers wanting to know why the belts were standing and no coal coming out

and we couldn't stop laughing. I don't think the bosses
ever knew what really happened that day.

'Another incident one day as our set of men came outbye
was when we passed a lad coming inbye who shouted to
us, "Got some bad news, lads – Kennedy's been shot."
Now it so happened we had an overman called Kennedy
who was temperamental and got upset when things went
wrong at the coalface. He would shout and scream and
jump on his hat and things like that. We thought it was
him and one of me marras said, "Whey, Aa knaa owld
Kennedy was bad-tempered, but thaas naa call to shoot the
man for that, thaa'll be war on aboot this." When we got
to the shaft bottom we learned it wasn't our Kennedy who
had been shot but the American President John F.
Kennedy.'

Superstition ranked high amongst the folk in the old mining
communities. This is not surprising considering the lives they led.
Death and disasters were common occurrences. If a husband or
any member of the family was killed, their anniversary would be
honoured and none of the family would work that day ever again.
Many of the miners would never work the day before or after a
holiday and certainly accidents occurred more often at these times
because it was natural prior to a break to hurry and finish work
as quickly as possible and head for home ignoring the safety
aspect. After a holiday men might be not as vigilant and
conditions they worked in may have deteriorated after the pit
stand-down.

Tales of ghosts, lights and figures appearing in old workings
were commonplace. It is easy when alone four or six miles
underground in a cramped roadway, far away from anyone, for
the mind to play tricks. Not a sound save the drip, drip of roof
water and the eerie glow of your cap lamp, then maybe a trickle
of loose stones pattering to the floor behind you. None of the
older miners would go to the pit if a black cat crossed their path,
nor would they when coming face to face with a woman in the

The three shafts at Ellington in the 1960s sunk between 1909 and 1913. No 3 upcast in the foreground was used for man-riding and later converted from a cage to a Lift Shaft. Nos 2 and 1 downcast shafts were mainly for coal drawing.

dead of night, that was a sign of death. Breaking a mirror or tea bottle were considered bad omens, as was snapping a bootlace before leaving for work and there are many more.

Friday the 13th was thought to be a day to be wary of, as Jack Tubby told me:

'It was during 1965 and one particular Friday which was the 13th day of the month that an incident occurred down Lynemouth Colliery. There were a fair number of men employed there from the local fishing village of Newbiggin and the fisher folk were by nature very superstitious. It was dayshift and the men underground were travelling inbye on the man-set for about four miles to the inbye station. The banter was as usual about the coming Saturday football involving Newcastle and Sunderland, while some men played cards. Nothing to suggest what was to follow. As the men were embarking at the station there was an almighty yell. Men came rushing back from the front and got back on the man-set.

'I wondered what the hell was going on. Was it an accident or confrontation of some kind? On reaching the inbye side of the train there was the problem staring me in the face. Some joker had nailed a pig's head to a wooden prop. Now one thing that is never mentioned on Friday the 13th is any reference to a pig to anyone superstitious and certainly not to fisherfolk, and here was one right in front of their eyes. I was one of four undermanagers that day and no amount of talking could persuade the Newbiggin men to go to their work past the pig's head. They had seen it and were going home away from the dangers of the pit, and off they went back outbye on the man-set. Men had to be redeployed to continue production. The manager was not happy when told of what happened but as production wasn't affected he had a wry old smile to himself.'

Chapter 5

Pony Patter

Horses and ponies have been used down the pits to haul coal, roof supports and mining machinery for some 300 years. The county's coal owners obtained their horses and ponies from dealers who bought them in from Galloway, Wales, the New Forest, the Dales, Ireland, America and Scandinavia.

Their importance to the working of the mine and the sheer disregard of human life by the Victorian coal owners can best be seen in this incident when a pony was killed by a runaway tub full of coal. The mine manager called a young putter lad to his office and asked for an explanation about the death of his pony. The putter lad said: 'It could have been me, sir, that died. I had to dive out of the way. There was no time to do anything about it.' The manager replied: 'That may be so, boy, but one pony costs the company £10 to buy and there are plenty boys like you who are seeking work.'

Things changed for the boys and the ponies after the 1911 mine safety regulations came into practice. No longer did the ponies have to work long hours, often with open wounds and sores. Now they had well-lit stables and proper feed and the regular attention of a vet and horsekeeper.

In his book *Gallowa – the Story of a Pit Pony*, Mike Kirkup wrote:

I had started my pit career at the bottom of the Bothal Pit shaft, coupling tubs into sets of four that were hauled away down the Bensham Drift. But when the back-shift overman suggested that I might like to work with a pony I jumped at the chance, although I had never even been close up to any kind of animal, except the pigs my father kept on waste land at his allotment near the Ashington Hospital. The idea of having a live horse for a marra (mate) excited me.

When the pits had first started here in the middle of the 19th century, most of the ponies used by North-East collieries were bred on farms in Ayrshire, Scotland. One particular breed came from an area called Galloway, and the Ashington ponies soon became known as 'Gallowas'. Because of their height, they were ideal for hauling coal and timber along the low, narrow workings which riddled the pit like rabbit warrens.

Being an ex-miner, I have many tales in store concerning pit ponies. At Ellington Colliery one Foreshift in 1957, which means starting work at midnight, I had been timber-leading on 'O' conveyor with a gallowa called Dai and was on my way outbye at the end of my shift to the shaft stables by the North West Diamond back-place. My carbide lamp spluttered and eventually went out leaving me and the gallowa in the inky blackness. I had no spare carbide to fill my lamp and give me a light. It was total darkness, claustrophobic and scared me to death.

It was Dai who saved the day when I climbed on his back and he took me out a mile to the diesel roadway and the lights, only stopping for me to open air doors across our route and wait for me to clamber up on to his back. He could have wandered into an old working but he didn't and we became the best of partners with a special bond forged between us. That was a lesson learned the hard way and I always carried spare water and carbide after that.

One pony I had down the yard seam was Steve, who had a wall

*Ellington miner Bob Pegg leads grey Welsh Mountain pony Flax who is
hauling out reclaimed pit props from a worked out roadway in 1994,
shortly before all Ellington's ponies were taken out of the mine. On the left,
Ian Miller guides black Fell Cob Tom inbye for another load to haul.*

eye and most of the time he was fine to work with but on odd
occasions he would take the bit between his teeth and run flat out
with a load of timma. Before he hit the turn into the Barrier
Roadway and tram and props flew everywhere, I would fling
myself off the limmas and under the belt. Why he did this I never
knew. Bob Bell said: 'One incident at North Seaton Aa'll never
forget is when the powny I was working at 32 Flat refused to

move and nee amount of coaxing could get him going. Shortly after, the roof collapsed right where we would have been. Aa think they have a sixth sense, thank goodness for that.'

There are many humorous and sad stories heard from other pits, but probably the classic one is this: 'There was one bloke at the pit who loved his pony and one day brought him out to the stable at the end of his shift. He said to the pony, "Aa'll see yi the morrow, bonny lad," and then left to take the cage ti bank. Just before he got in the cage he turned round and raced back to the stables to his pony and whispered in his ear, "No, that's not true what Aa said aboot Aa'll see yi the morrow, 'cos Aa forgot Aa've got a rest day the morn." '

Keith Adams, farrier-horsekeeper at Ellington Colliery for 23 years, takes up the story:

> 'It was serious business looking after Ellington's ponies and I was on call 24 hours a day, but there were many humorous times and I loved every minute of it. They all had their own natures, just like us. I washed them, fed them, groomed them, mucked them out, and shoed them, as well as seeing to their health when necessary.
>
> 'There was one I remember, little Able, who came from the pit at Sacriston, he was so small at 9.3 hands I had to get down on my knees to shoe him. Then there was Rap, who when he got new shoes would kick them off against a door as soon as he left the underground stable. Simon was a real card, he loved to roll in the dust but the trouble was he was so big he sometimes couldn't get up again. This meant we were called out to go inbye with chains to winch him back on to his feet. Bob 1, so called as we had two named Bob, hated water and he would rather walk on rails than get his feet wet. A lovely little grey called Spring was frightened of the dark and hated being left alone and if he was would whinny and squeal till his handler came. Sadly he got loose one shift down the 4th North and was hit by a loco which took his foot clean off and he had to be put

Ellington Colliery's blacksmith farrier Keith Adams replaced ponies' shoes usually every three weeks. Here he is at his work in the underground stables giving New Forest pony Regal a set of new shoes.

down there and then. Then there was Toby, who was scared of shadows or sudden draughts of wind. The handlers loved their ponies and would bring bags of food in for them at Christmas – some of the stuff would not have been out of place at a Mayor's banquet and no good for the ponies. The lads loved a bit fun and one Christmas I remember going in the stables to find some of the ponies' stalls all decorated.'

Lynemouth Colliery horsekeeper Bob Gilholm and pony Roy who won Best Pit Pony at the 1947 Morpeth County Agricultural Show.

Charlie Farrell of Bedlington, who had previously worked at Cambois and Bedlington 'A' pits until they closed in the 1960s, looked back on his time at Ellington during the 1980s:

'I took Bob 2 one day inbye to the Grangemoor district. At the end of the shift we were coming outbye to the stables when we passed a lad called Billy Johnson who worked on the belts and he fed the gallowa some bait he had left. The same thing happened for four days and on the fifth we came to the place where Billy worked but he wasn't there. Bob 2 stopped and refused to move. After ten minutes Billy turned up, fed the gallowa with his bait and only then would he move on to the stables. This happened many times and Bob 2 would not pass that spot until he had been fed!'

Chapter 6

Bevin Boys

I t was almost 70 years ago, during the Second World War, that 48,000 young men were shocked to learn that instead of fighting for King and Country they were being conscripted to work as miners under a scheme set up by Ernest Bevin, the War Cabinet's Minister of Labour and National Service. They were chosen by ballot to fuel the war effort and replace the masses of experienced miners who were fighting abroad in the armed forces. Many of these Bevin Boys – as they were ever after known – came to the pits of Northumberland.

Now in their eighties, men like Sid Stowe originally from Oxfordshire, John Marshall from London, Philip Renforth from Newcastle, Ron Jones from Birkenhead, Bob Hughes of Dunoon and Ron Simpson from North Wales were only a few of the thousands forced to become Bevin Boys and very much against their will. It took a long time to recognise the service that the Bevin Boys gave to their country. At the millennium they were included in the yearly November march past at the Cenotaph and an official badge has been awarded to the survivors after all this time.

Sid Stowe remembers being sent north as a 17-year-old: 'I thought I was coming to the end of the world, before then I had never been further than Oxford station. It was terrible 'cos I was all geared up to go to war and instead ended up down Ellington pit.' He recalls finding lodgings with 'wonderful people', the Cox and Wilson families from Pont Street in Ashington, and integrated so well with the locals that he decided to settle in the town after the war ended.

Ron Jones was training to be a plumber before being called up to the pits and had wanted to join the Navy with his best friend, but instead found himself doing war service down Newbiggin Colliery: 'At first it broke my heart. But I knuckled down, did my best, and made some good friends, finding the local pit folk so kind and friendly towards me.' After his service ended he returned home for five months but was unable to find employment. He contacted the pit manager back in Newbiggin and asked for his old job back. Ron married his wife Lillian and stayed at Newbiggin for 23 years until the pit closed, then worked at Woodhorn and finally Ellington.

Bevin Boys Sid Stowe, Phil Renforth, Ron Jones and John Marshall reminisce about their days during and after the war years working underground in the local pits. Pictured in the pit yard at Woodhorn Colliery Museum in June 2007 at a special gathering of Northumberland Bevin Boys.

John Marshall looks back and thinks of the bitterness he felt at being called up as a Bevin Boy: 'I was forced to work in the pits before being allowed to join the Paras. During the war I corresponded with a Northern lass and came back to marry pen pal Connie and settle in Ellington where we still live today.'

Philip Renforth wanted to join the Navy but ended up at Woodhorn Colliery, where he stayed after the war until the pit closed in 1981 and he transferred to Lynemouth where he still lives with his memories: 'It was hard for us them days as some folk thought we were conscientious objectors or cowards for going down the pit rather than fight for our country, but we simply had no choice. My first job was on the trot road before graduating to timber leader and then on to the coalface where I was a cutterman and from there on to the arc shearers before becoming an Official. Being a Bevin Boy changed my life forever, but I cannot grumble because I ended up with a good life as the work was always there, and I have been married 56 years, with me and the missus bringing up our family here.'

Bob Hughes left the cosy seaside village of Dunoon to enter the harsh world of mining at Newcastle's Seaton Burn Colliery:

'It was hard for me at 18 years old and I often worked through the night. Trouble was that I was lodging in Newcastle and the last bus to get me to the pit at Seaton Burn was 10.30 pm. So when I got there it meant hanging around for another two to three hours to start my shift. It got too much for me and I went AWOL and returned home. I was prepared to take the consequences even if it meant a spell in Durham Gaol. My father persuaded me to go back and tell the court my circumstances. The Newcastle magistrate asked if I would return to the colliery if they changed my digs. I agreed and was found lodgings at Longbenton where I could get a late bus right outside my door. After a while I managed a transfer to the Navy. Being a miner was not for me, it was a hard life and I have

*Depending which pit a miner worked at there was either a wash or a
water hose to clean the sweat, sleck and coal dust from his gallowa
after a hard day's graft underground.*

every admiration for the local men who braved it year in
and year out.'

Tom Cowan, who worked at Broomhill Colliery, remembers the
Bevin Boys who were sent to work there:

'The Bevin Boys had no hostel to be put up at and so
stayed with mining families. Some of them were okay but
pitwork was alien to them and there were lots of accidents.
The ones housed at East Chevington were appalled at the
old-style netties and one wouldn't even use them because
of the stink, instead he used an old shed as a toilet.
He went back home to Ilford after four days and never
came back, saying he couldn't live in such primitive
conditions.'

Ron Simpson rolled back the years when he told me:

'It was the morning of my 18th birthday, 1944, when I was called up for the pits. I had never been down a pit in my life and ended up at Choppington. Me and a lad named Jack, who we nicknamed Gracie 'cos he was from Rochdale like the singer Gracie Fields. After only four weeks' training at Bedlington and Cramlington, we lodged with Philip and Martha Smith at Eastgate. I remember having to travel with her to Bedlington so we could register our ration books so she could obtain our food rations. As miners we could claim a higher meat ration then. There were two pits at Choppington, the 'A' and the 'B', Jack was sent to 'A' pit and me the 'B' pit.

'Mr Oakley the manager told me to report for work in Foreshift the next day at 2 am. I had to buy my own carbide lamp from the local store which cost £1 10s 0d. My first job was on rope haulage with a wage of just £2 7s 0d. I had

An early picture of Choppington colliery which by the end of the Second World War was noted for having the smelliest pit heap around the county. It glowed in the dark and gave off noxious fumes and if the wind was in the right direction public transport users knew they were approaching Choppington long before they ever saw the place.

to pay £1 10s 0d for lodgings and was always broke by Monday morning. After a while I was moved inbye towards the coal face and the putters landing where lads worked their ponies pulling empty tubs into the coal fillers and the full tubs out. The putter lads were on piece work and getting good money. I remember the day the manager came inbye and I asked him if there was any chance of me becoming a putter. "Are ye stupid or just mad, bonny lad?" he said. "Neither," I replied, "just broke!"

'Next quarter came my turn to go putting and I went to the stables to yoke my pony. His name was Orth and one of my putter marras another Bevin Boy called Ramsay. It was hard work and when working in the Busty Seam very wet, no one liked those conditions. The skill of a putter was keeping the tub with its half ton load of coal on the rails. If not it meant unhooking the pony and lifting the tub back on. The putters and fillers accepted me as one of their own. As long as I put all my effort into the job they never treated me as an outsider, I was one of their marras. Now I was earning £12 and very happy.

'One day I was putting in the Busty Seam for a filler called Jack Dent when the roof collapsed behind us. We managed to get out, but had to leave the pony till later while the fall was cleared. I remember two men were killed in the three years I served there between 1944 and 1947. Here is a song we putters used to sing that helped to make light of the conditions we worked in:

As Aa was going inbye one mornin'
Just alang the rolleyway,
Aa met a putter and he was singin'
"Aa'm off the bloody way",
He asked "Can yi give me a lift,
'Cos me arse is red and sore,
And when this bloody tub's back on,
Aa will put nee bloody more." '

Chapter 7

Moving On

After the end of the Second World War in 1945, a Labour government was elected to take the country forward. With that came talks to nationalise the coal industry and this happened at vesting day on 1st January 1947 when the country's pits became publicly owned. A huge sense of relief and optimism grew in the hearts of the British miners. The old regime of the coal owners, after years of dictating terms and conditions in their privately-owned pits, was finally put out to grass, but it cost the country £66 million in compensation to pay off the many coal companies involved and then start a major investment programme in the mines.

Alf Goodall has lived in the pit village of Lynemouth from when he was five years old and remembers the day:

'I started work at Ellington Colliery in 1935 and after the usual breaking in work on the heapstead was accepted as an apprentice electrician. The Ashington Coal Company had a reputation for being hard but fair employers, and this I found was about right as I had every opportunity to benefit from their education opportunities and a good grounding in electrics at their other company pits of Ashington, Woodhorn, Linton and Lynemouth. After the

A spate of new Lodge banners came into being after the country's pits were nationalised in 1947, with many others being refurbished to mark a new beginning in the coal industry.

pits were nationalised in 1947 there was little difference in the management as most of the old staff still kept their jobs. What did change was the National Coal Board invested huge sums to improve safety and mechanisation in the pits to ensure their long term future.'

Old Mathew Tait, who started work as a trapper boy, was amazed at the change in working practices at some of the bigger pits like Lynemouth and Ellington after nationalisation:

'Now it was the time of the big mechanical hewers – continuous miners, ploughs, shearers and trepanners which both cut and loaded the coal – diesel locomotives which pulled out 100 tons of coal a time to the shaft

The old style of setting roof supports on a Longwall face.
My old marra Dave Brown in the Yard Seam at Ellington Colliery
builds up hardwood chocks.

bottom. On the surface, tower winders, coal preparation plants and computerisation would now see the miner as a highly skilled worker.'

The 1960s may have been the rock and roll years of music but they are sadly also remembered for the demise of many of the smaller and low production pits in the coalfield. The Government now looked to other economic sources of power such as oil residue. The major pits with extensive reserves of coal were being upgraded both on the surface and underground to cope with increased output. The men from the old pits had the choice of moving on to long-life pits in the area or to the Midlands coalfields, and some left to work at local factories. For the older miners there were fewer options and many never worked again.

Hi-tech modern machinery was brought in, trialled and, if successful, installed at many of the expected long-life pits from the 1950s. Seen here are Ellington miners Alec Spowart at the controls and Joe Davison trialling a new roof support machine.

There was little opposition from the unions about closures at that time. Some lads enjoyed the new experience of moving on and others didn't.

George Graham of Westerhope followed pitwork after his home pit at Prestwick closed in 1965. First to East Walbottle Colliery and on to North Walbottle before it closed in 1968, after which he transferred to Dudley. After a short time out of the pits he missed the camaraderie and returned in 1975 to work at Brenkley, where he became Union Secretary until the pit closed in 1985 after the miners' strike.

At the beginning of the 1960s, membership of the National Union of Miners stood at 700,000, which by the end of the decade was reduced to 290,000 with the miners way down in the wages league. Of the 70 Northumberland pits open when the pits were nationalised in 1947 and employing 36,254 men, by 1970 only 16 pits were left with a total workforce of 9,600 men.

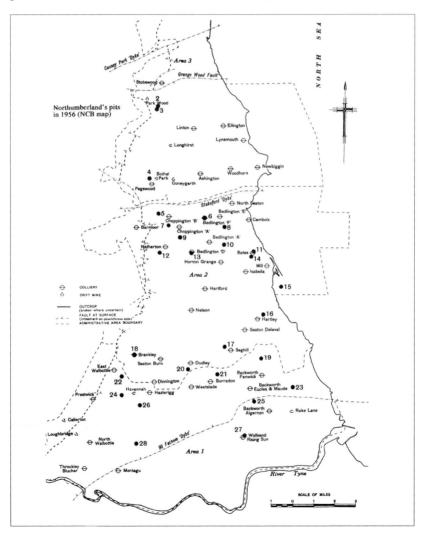

Map of 1956 from Blood on the Coal.

Northumberland Colliery Closures 1960s
with their lifespan as from original sinking

Callerton Drift	1950–1960
Dinnington	1867–1960
Maude	1862–1960
Morewood	pre1910–1960
Seaton Delaval	1838–1960
Broomhill	1849–1961
Hartford	1858–1961
North Seaton	1859–1961
West Wylam	1866–1961
Barmoor	1882–1962
Bedlington 'E'	1859–1962
Horton Grange	1853–1962
Seghill	1824–1962
Loughbridge	1947–1963
Seaton Burn	1838–1963
Hazelrigg	1892–1964
Bedlington 'F'	1854–1965
Prestwick	1771–1965
Stobswood	1875–1965
Algernon	pre1900–1966
Choppington 'A'	1857–1966
Choppington 'B'	1857–1966
East Walbottle	1908–1966
Hauxley	1926–1966
Isabella (Cowpen)	1848–1966
Weetslade	1903–1966
Newbiggin	1908–1967
Bedlington 'D'	1854–1968
Cambois	1863–1968
Linton	1894–1968
North Walbottle	pre1890–1968
Longhirst	1956–1969
Mill	1885–1969
Pegswood	1872–1969
Rising Sun	1784–1969

When I spoke to Bob Bell he told me:

'When North Seaton Colliery closed in 1961 Aa transferred to Choppington and worked on a shearer face which was okay until it closed and from there moved to Linton Colliery, but didn't like that as the overman was a funny bugger and had a strange way of drawing cavils. Instead of letting men draw their next quarter workplace from a hat, he hoyed the slips of paper on the floor saying sarcastically, "There yi are, tek your pick." That was nee way to treat men and there was a big bust up. From there I went to Lynemouth working on the West side and that was nay better. So Aa just up and left the pits. It's never the same when you leave your home pit and all the lads you grew up with; there's never that closeness or camaraderie with folk from other pit villages.'

Jimmy Swan also recalls transferring to Choppington:

'I got on well with the Choppington men, but was then sent to Barmoor which was just a dinky pit having just three deputies, four drawers, five stonemen and ten coal fillers. They spoke a different language to us, calling stones "clemmies" and things like that. Back I went to Choppington until it closed and then to Ellington and what a difference with the high roadways and new machines working coal seams six and eight feet high. The only drawback here was that we newcomers from other pits couldn't get on piece work 'cos there was a waiting list of Ellington men for this and no way would the union allow us to get ahead of the queue.'

Kit Miller spoke to me about the men who came to Longhirst from Lynemouth Colliery after the 1966 fire:

'It was a big move for the men who had been used to

working in high seams with modern machinery. Here some of them had to work on the coal face with barely room to move. Until they became used to it many were laid idle with beat knees and aches and pains caused by the conditions, so this ailment was nicknamed the "Longhirst Syndrome" by local doctors.'

Geordie Hume mentioned his move from Lynemouth Colliery after the fire there in 1966:

'I was sent to Ashington Colliery and what a difference that was. They were working a Longwall Shearer on a low coalface – it just looked like a flatiron moving along. I was amazed 'cos Lynemouth were working Continuous Miners in 8 to10 foot high seams. Aa said to the men, this is going back to the old days, the little coal you're getting is only what Lynemouth would leave behind and not bother with. Aa stuck it out there before moving to Ellington and then back to Lynemouth. Some of the Ashington men moved here when their pit closed and said to me: "Geordie, Aa can see what yi mean noo aboot the way we got the coal at Ashington." '

Electrician Mel Brown did his face training at Hauxley:

'It was a wet pit with Longwall faces like my home pit at Shilbottle where I served my apprenticeship. When I transferred to Ashington Area Workshops, what a difference that was. I missed the humour and the crack, it was like being in a prisoner of war camp. I worked on stripping down JCM miners and that was okay, but the foreman was in an office overhead looking down on everyone working on the shop floor. It was a bit intimidating and lots of the other lads felt the same. I got the opportunity of a job at Lynemouth Colliery and jumped at the chance and stayed there until Lynemouth

Machine Instructor Joss Hanson, centre, in 1982 showing new recruits G. Edmundson and Herron, who were transferred from pits around the area, how to operate the handles of a Joy Loader in the Training Gallery at Ellington Colliery.

and Ellington combined in 1983 and was then officially an Ellington employee. What a change that was working there with hi-tech machinery and thick coal seams with loads of head-room. It made me realise how far things had progressed from the primitive methods of coal getting at the hand filling pits like Shilbottle and Hauxley.'

When a pit closes it often sounds the death knell of the village that was built, grew and prospered in the shadows of the pit headgear. The villages of Widdrington, North Seaton, East Chevington, Radcliffe, Longhirst, Stobswood and many more throughout the county have suffered this fate. Homesteads have been demolished for opencast coal mining or for new developments, resulting in a loss of identity when mining families moved into new communities.

Joss Hanson, who schooled many of the transferred miners as instructor in Ellington's purpose-made Surface Mechanisation Training Gallery, had this to say:

'The lads that came from all over the coalfield had never seen big machines like the JCMS. They had come from pits where hand-filling in three-foot seams was still going on so it was a real eye-opener for them. Ellington was fast becoming a super long-life pit with giant machines that ripped the coal out in eight foot six

The East Coast Colliery of Hauxley had a short lifespan of 40 years
with its best years being in the 1950s producing 180,000 tons per annum
with a workforce of 564. The pit was closed by the NCB on
26th November 1966.

high seams. Some of the transferred lads were quaking in their shoes when they first handled the giant machines, but it was my job to instruct them and that's what I did for a whole week to start with and they soon learnt the ropes. Robert Bell, Tommy Hindmarsh and Ian Page transferred from Whittle and said, "We have never seen owt like this before, but it's amazing how much we learnt in a week." '

Chapter 8

Pit People's Picnic

When **Thomas Hepburn created** the Northumberland & Durham Miners' Union in 1831, it was the signal for each separate pit to show their allegiance to the union by forming a lodge, which then commissioned a huge bright banner made with the name of the colliery and usually a scene showing some aspect of mining history. Displaying their banners was one way of showing miners' identity and solidarity when engaged in disputes with coal owners. They were held proudly aloft when marching to organised meetings. Eventually most collieries in the county had their highly decorated banner.

It was the return of a Labour government in 1945 and the nationalisation of the coal industry in 1947 that saw an upsurge of new banners to celebrate these events. The first ever Northumberland Miners' Picnic was held at Pollys Folly, near Shankhouse in 1866, three years after Northumberland formed their separate union. The picnic has continued right up to the present day save for the war years and a rally during the '84 strike.

Morpeth and Bedlington were favourite venues as a day out for

Ɛɖɖ̄/ᴎGTON

NATIONAL UNION OF MINEWORKERS.
NORTHUMBERLAND AREA.

RESULT OF

BAND (MARCH) CONTESTS

AT

ANNUAL PICNIC

HELD AT

BEDLINGTON

ON

SATURDAY, 9th JUNE, 1973

ALSO

AWARDS and REMARKS of the ADJUDICATOR

(Mr. A. Chappell, Leicester)

Band contest winners and entries at Bedlington Annual Picnic,
9th June 1973.

the miners and their families to celebrate, this was their day. Thousands of folk turned out to watch the colliery brass bands marching and competing. On stage at the picnic field were the miners' leaders and prominent Labour Party members who

85

With a police escort the Miners' Union leaders head the Picnic Parade and move off down Ashington main street towards their base at Woodhorn Colliery Museum for the day's celebrations.

delivered stirring speeches on the state of the coal industry and the country. A Picnic Queen and attendants were paraded on a float, and food and ice cream stalls were installed in prominent positions, plus fairground shows to keep the young folk happy. And not least, the local pubs opened their doors early to thirsty miners and bandsmen.

After the majority of the pits in the county closed, the later Miners' Picnics were held at Ashington, with a parade from the main street and down to Woodhorn Colliery Museum. Although the last of the pits in the county closed in 2005 a speakers' session at Ashington on a Friday night in June, followed the next day by a Miners' Memorial Service in Ashington parish church and afternoon events at the revamped Woodhorn Museum and County Archives Centre are today's events and are still well supported.

Chapter 9

Disasters

The annals of coalmining history are filled with stories of disasters. In our county, New Hartley, Wallsend, Montague, Burradon and Woodhorn Collieries are only some where major tragedies have occurred. A disaster is often declared in official terms when a certain number of men have lost their lives in a major incident, but to any pit family the loss of one breadwinner is a disaster in itself. The closing down of a viable colliery for whatever reason can also be termed a disaster as this affected the many miners and their community who relied on the pit for their livelihood.

Bob Bell looks back on two disasters:

'I was working at North Seaton during the 1950s with a set of drawers taking out roof supports when Andy Eastlake was killed. He was chopping oot a prop when the roof collapsed and he was buried under a big stone. Andy had nee chance, he was dead when we got him out. When North Seaton closed I moved to Choppington and again a lad was killed by a faulty Dowty Prop on the shearer face I was working on. I think his name was Moffat.'

Alf Goodall told me of his early experiences at Ellington Colliery in 1936:

'While working on bank came an event that I can still picture today when a bloke called Steadman was killed, crushed by tubs; and the same thing happened to young Raymond Barker. I was 15 years old and Raymond was a friend of mine and a Chapel lad who lived down the same street. Being just a lad then and new to the pit, it was a terrible sight and an experience that haunted me for years. Folk never think that many fatal pit accidents happen on the surface, but they do.'

Alex Anderson tells of his experiences in the 1940s at Woodhorn after becoming District Overman:

'If there was an accident inbye, sometimes it meant the patient had to be stretchered two miles to the shaft bottom by relays of men. A serious case would have to be injected with morphine and I had qualified to do that. One case I attended was when a young haulage lad had his leg torn off in a horrific accident. During my time as overman during the 1940s and '50s one of my fellow overmen was killed and I was asked to inform his wife and family and I found that very hard, but somebody had to do it.'

Miners never forget incidents like these and Kit Miller recalls vividly one such occasion at Longhirst:

'A lad called Freeman was coming off the face and had his arm trapped on the tail end of the belt when the stone canch eased down. Luckily some of the weight was taken by the cutter, but he was in agony and no way could we get him out. Each time we moved the tail end he screamed blue murder and by then the doctor was with us. He wanted to give him a high dose of morphine but I said no 'cos we needed the man's help to tell us if we were doing the right thing.

'Eventually we had to lower the tail-end by howking

The Woodhorn Colliery explosion of 13th August 1916 claimed the lives of 13 men. The impressive mining memorial to those killed was sited in the Hirst Flower Park, Ashington, in 1923 and later moved to its present home when Woodhorn became a Colliery Museum in 1989.

underneath it to release him and the doctor stayed with us for four hours. He was amazed at how we got him out. The lad had serious injuries to his arm and never came back to work at the pit again.'

Jack Tubby told me of a day in his life he will never forget:

'Monday, 6th November 1950, my dad and I were both in backshift due to go down the pit at West Sleekburn Colliery at 9 am. We donned our pit clothes and said our goodbyes to mother and my two younger brothers. As we walked the short path to the pit we spoke little; all we needed to say was done at breakfast. I always thought my dad had never forgiven me for following him into mining. He and mother had scrimped and saved to send me to private school which didn't work out and here I was following my marras who were now earning. We collected our lamps and said our goodbyes and promised not to wait for each other at lowse and just get straight back home.

'At the shaft bottom I was deployed by the Head Rolleywayman to the trot road tub haulage a mile inbye from the shaft to receive empty tubs from the shaft and to send full tubs out. My dad was working on a coalface just outbye from where I was. He was a drawer, moving the belt system and roof supports. Working down the pit was dangerous, but I knew his job was particularly hazardous.

'After about two hours the whole haulage system stopped, so after half an hour I decided to walk outbye to the next transfer point to see what the problem was but none of the lads working there knew. Then Mr Bill Birnie, the head of all the pit's haulage system, appeared and told me to put my jacket on and follow him to the shaft. This was strange as it was nowhere near the end of my shift and so I asked him why. He said there had been an accident on the coalface where my dad worked and he was involved.

'I couldn't believe it and asked him for more information. At first he wouldn't, but seeing I wasn't prepared to move from the spot until he told me how serious dad's accident was, he did finally tell me. He said, "Are you sure you want to know? I would rather you came to bank with me now." I replied, "If it's bad news, I would rather you told me now." Little did I realise how bad the news would be, still thinking it was an accident and no more than that.

'Bill Birnie said, "Well son, I haven't had to do this before. I feel terrible and so sorry to have to tell you that your dad is dead, killed by a fall of stone." I felt cold inside, this couldn't be true surely, then I looked at Bill's face.

'I have no recollection of the walk outbye or what happened to my pit lamp but as Bill walked me home and left me at our house my thoughts were with mam. Would I be the one to tell her and what could I say? I'll never forget the look on her face as she stood at the door and I knew then she had been told of the tragic news. Her brother Jack had hurried out of the pit to tell her and I was grateful for that. News travels fast in a mining village and our house was full of neighbours wanting to help in any way they could. It was some time before I had the chance to change out of my pit clothes and bathe by the fire.

'After the funeral we learned exactly what my father's life had been worth. The conditions of any sum of money being offered to mam was for the pit to continue working on the day he was killed. It was custom at West Sleekburn Colliery for this to be the accepted procedure and mam was given £200. Was that all my dad's life was worth? Two hundred bloody pounds? Life went on but it was difficult with only my own paltry wage to keep my mam, me and my two school-age brothers, Bill and Alan.

'Losing dad made me more determined to carry on studying to make something out of myself. One thing to come out of our disaster was the fact I had to learn to grow up quick and shoulder the responsibilities that a family without a dad had to face.'

Nance Walton was one of eight children left without a father on a fateful day at Woodhorn in 1916. Quoting what her mother had told her, she spoke of the tragedy to Mike Kirkup:

'A polis called to see my mother at seven o'clock on a Sunda' mornin'. She asked him what was wrong. "I wouldn't likely know that Mrs Walton, but you'd better get yourself down to the colliery, that's all I was told to tell you. I know nothing else."

'And with that he took his leave and was gone. Leaving me mother wondering and worrying what exactly had happened and if me dad, Ned Walton, was involved at all. But she got mesel' ready and set off for the colliery. It wasn't all that much of a distance to Woodhorn 'cos we were just living in Chestnut Street at the time, second block, that is.

A typical underground accident scene is created here on the surface as a Deputy attends to an injured miner in a National Coal Board First Aid Competition at Newcastle in 1961.

'"When I got to the colliery," mother said, "the gates they were shut and there was a big crowd there already, mostly women. Just then I heard the clanging bell of a fire engine. It was the Northumberland Rescue Brigade. A man in a trilby hat opened the gates shouting: "Stand back! Stand back there! It's a man's job to get these men out of the pit."

'And an old woman standing next to me said under her breath: "Aye, and it's a woman's job to bury them." She knew, you see. This wasn't the first time she'd had to bury one of her kinfolk. But she was right. There were thirteen men killed that morning in a gas explosion; twelve of them were married and left big families – me mother had eight bairns, so we were the hardest hit.'

Drawers at work in a low seam on a Longwall face, pulling over the belt into a new track as the coalface advances.

Lynemouth pit on fire

The year 1966 is one that will never be forgotten for many reasons. It was the year that England won the Jules Rimet Football World Cup, and then, after the euphoria, came sadness with the disaster that shocked the nation at Aberfan, a mining village in Wales when a giant slag heap collapsed and engulfed the local school with the loss of many lives. Then, 300 years after the great fire of London in 1666, came the great fire of Lynemouth.

Newcastle Journal
15 November 1966

An underground fire which has been smouldering for two years flared up yesterday.

Miners were told to go home and an NCB spokesman said last night it was uncertain when the men would be allowed back. The area was sealed off in 1964 after the fire was discovered. Later that year experts fought it with chemicals.

Eric Dodds looks back:

'I was onsetting at that time on the Main Seam Level at No 1 Shaft which was the upcast shaft. One day I was on duty with Les Taylor, a Newbiggin lad. Foul smoke and fumes were billowing out from the fire. On this occasion bottles of nitrogen used to combat the fire were coming down in trams and tubs. I needed to enter the cage to pull out a tub but was overcome by the fumes and collapsed. I was taken to the Medical Centre by Joe Blandford and given oxygen. Within minutes Big Tom Smith the pit manager arrived all concerned to see how I was. When I eventually recovered I went back down the shaft to my place of work.'

John Freeman, Under-Manager at Lynemouth, remembers the last days:

'The fire was now virtually out of control; all the measures taken to prevent oxygen reaching the source had failed. I was Duty Senior Official in the Backshift on the last day in November 1966. During the immediate Foreshift, a team of rescue brigade men had tried to reach a certain area of the fire. Smoke was so dense that the team were using a rope to hold onto. The story is that one of the team stumbled and lost the mouthpiece of his breathing apparatus. Fortunately he recovered it, otherwise it could have had tragic consequences.

'At about 8.30 am, a Senior Inspector of Mines and Quarries, myself and a skeleton crew entered the mine and made our way to the North Landing taking two diesel locos with us. From there we made our way to a designated 'Fresh Air Base' near Panel B district and the top of No 3 staple. I will never forget the scene that greeted us. The smoke was so thick that I could not see my marras only twelve feet away across the Panel B entrance. The heat was terrific and causing the strata to move. Then the telephone rang, it was Billy Waters the Deputy Manager with a message to round up everyone and get them to the North Landing. The pit was to be evacuated. Soon after I rang Billy to tell him we were leaving for the shaft bottom when the next piece of drama unfolded. Billy said that two pump men in my district Panel C, North 2nd East could not be contacted. It was now up to me and I asked for a volunteer to accompany me and without hesitation up stepped young Frazer Patton, a mining student.

'We began the long trek into my district with Frazer travelling one intake roadway and me the other. At 62-yard intervals we could see each other and stayed in contact that way. Eventually we reached the point where the men should have been, but there was no sign until a short search found them having their bait in a quiet back-place. I advised them of the danger and the need for urgent exit before phoning Billy to tell him we had found the men. His words still ring in my ears: "John, get to hell out of there NOW."

'Having worked closely with Billy, I recognised the urgency in his voice and with the others quickly travelled out on the conveyor belts to the North Landing, switching off the electrics as I went. Boy was I pleased we had decided to bring a spare locomotive and evacuate to the shaft bottom. My last act at Lynemouth Colliery was to work with a team to lay hoses from the Lyne Burn and watch millions of gallons of water being pumped into our pit to quench the fire. I don't mind admitting that a lot of teardrops were mixed with the river water flowing down the shaft.'

Newcastle Journal
19 November 1966

Miners and firemen last night began the biggest pumping operation in mining history in an attempt to flood a showpiece colliery to check a fire 600 feet below the North Sea and a mile from the shore. The fire began in Lynemouth Colliery, Northumberland, the world's biggest undersea pit which produces a million tons of coal a year.

Newcastle Journal
22 November 1966

The battle to bring safety to Lynemouth and its neighbouring pits is now well under way. Dr W. Reid of the NCB said: The ventilation fan at Lynemouth that was pressing air and gas towards Ashington and Woodhorn pits has now been reversed; first reports are that the air is clearing. Water has been pumped into the pit at the rate of 7,200 gallons a minute.

Alan Young recorded:

'Thankfully, no lives were lost during those dark days and owing to a remarkable action by management the ponies underground who were in danger of being lost were saved with no time to spare. Almost a hundred ponies were housed underground in the shaft stables. This was in the return airway and the most vulnerable part of the mine. Conditions in the pit were deteriorating by the minute, noxious gases polluting the airways and it would not be long before any rescue attempt would be impossible.

'There was only one way to effect a rescue and give the ponies fresh air for an hour or so. That was to reverse the main fan on the surface. A major decision for anyone in authority to make as it is one of the most dangerous operations to be carried out in a coal mine due to change in air pressures and other factors, and here they had a pit on fire. Reverse the fan they did, then sent down teams of rescue men to remove ventilation doors and create a flow

Ongoing training and practice was always available and encouraged at the pithead for enthusiastic miners. This photo shows an individual Firefighting Competition at Ashington timberyard in 1964. Nothing as extensive as the Lynemouth disaster here, but still dangerous.

of air to the stables. Within hours the ponies were taken out of the pit and all survived but it was a close run thing.'

The flooding of the pit with over 200 million gallons of water meant that most of the 1,700 workforce were literally thrown out of work. The East Arterial District which contained the main transport links of the colliery and the lower winding horizon at both shafts were permanently flooded. Fortunately the Coal Board had plans even

After the fire, work began on driving a new drift to link up with the coal seams to the north and to connect with Lynemouth's sister pit of Ellington.

The flooded winding shafts at Lynemouth are shown here being demolished, no longer necessary when coal began to flow to bank from the new Bewick Belt Drift.

then for the reopening of the mine and men were sent to work at other local pits such as Ellington, Ashington, Woodhorn, Linton and Longhirst.

To the rescue

Specially trained miners from local collieries formed the Area Mines Rescue Team. Alex Anderson was one of those men and told me a fascinating story:

'I was an overman at Woodhorn in 1945 when the manager asked me to train to become a member of the Colliery Rescue Team. I accepted and training lasted for 14 weeks at Ashington Rescue Station before I qualified. Training was intense, working under simulated underground conditions wearing all the necessary breathing apparatus and equipment needed in any pit disaster. An aviary of canaries was kept there which we used in our exercises for detecting bad air. At one time, two were kept at each of the Ashington Coal Company pits for convenience in case of any incidents. There were lads in the team from different pits around the area and we did a day at all those pits to be familiar with their workings. I served 16 years with the team and received a certificate and gold medal at the end.

'I served on active duty at three underground fires and two disastrous explosions. The most traumatic of these was the 1947 explosion at the William Pit at Whitehaven in which 104 miners lost their lives. Rescue Teams worked in relays two hours at a time and our job was to locate stricken miners and if possible bring out anyone alive in the pit. We crawled over and counted dead bodies but never found any survivors. One day I shall never forget when we were coming out of the pit gates. We were not allowed to speak to the public about the incident but a woman pushed her way through to me and held up a photograph saying, "Have you seen this man, he's my husband – have you seen him?" I saw the desperate look on her face and felt so sorry for her and all I could do was shake my head.

'The people of Whitehaven were great with us and took us into their homes, even giving me a change of clothes when I first went. The lady I stayed with knowing the reason we were there never once mentioned that she had lost her husband in a pit accident the year before. Our team worked for five days until all the bodies had been located and families informed. The year of 2007 was a memorable time for me as the folk from Whitehaven asked me to lay a wreath at the Disaster Memorial to commemorate 60 years since the tragic event. It was a great honour for me as I am probably the only one on the rescue side surviving today.

'Weetslade Disaster in the 1950s was different in as much that we were based at home and travelled each day to the pit. Five men lost their lives and although we spent ten days there it was some time after that their bodies were brought out. The men had been working in a salvage area when there was an explosion reportedly started from a battery shuttle car, and falls of stone and bad air with rising water levels that prevented any rescue. There was also the danger of further explosion through water at the battery station. I believe they had to drive new roads through before they reached the area where the missing men worked and it was the following year before they were able to locate them and bring them out.'

Kit Miller, a member of the Rescue Brigade in the 1950s, remembers two incidents and told me:

'I was knocked out of bed early one winter morning by an NCB van driver who had a message for me to report to Ashington Colliery. He left and I had to bike to the pit from Widdrington. There was an underground fire with a belt and props blazing away which was controlled after a few hours. No one was hurt but a fair bit of damage which could have been much worse if it had not been spotted in time. The second alarm was at Woodhorn, an almost identical incident to the Ashington one and controlled in good time by men trained to deal with the situation.'

First aid in the county's pits was, at the beginning of the 20th century, mainly the job of Deputies and other officials who were in charge of districts underground. After the introduction of new safety legislation many were encouraged to join the St John Ambulance Brigade and held practice sessions at the coal companies' pithead locations. Often the local doctor would conduct the proceedings and test the men on their worthiness to give first aid. From the 1940s first aid was high on the agenda of the coal

*Mines Rescue Headquarters on Station Road, Ashington, next to
the council offices, showing a Fire Brigade inspection
during the 1920s.*

authorities, who employed highly skilled instructors at their training
centres for proposed young miners from the age of 15.

Alan Young looks back at his training:

'In 1951 I went to Ashington Training Centre for a 16-
week course that took the new young recruits who would
eventually go to work underground. The lads came from
many pits like Ashington, Lynemouth, Ellington, Linton,
Woodhorn, Newbiggin, Hauxley, Shilbottle, Pegswood
and many others. We were kitted out with all the pit gear
we would need. We all sat a basic exam to see if we were
able to do further study and put in groups according to our
ability. A lot of our time was spent on films and lectures
and on the surface before going down the pit. Our First
Aid Instructor was the well-known Wilf Dick. I joined the
First Aid Team and we entered local competitions. There
were four in a team and we all had jobs to do such as

*Bob Tweddle ready kitted out for a rescue exercise at
Stobswood Colliery in the 1950s.*

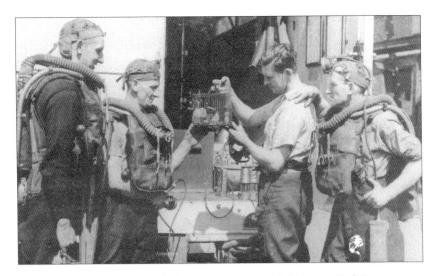

Checking out the canary before an exercise are Ashington Coal Company
Mines Rescue members: from the left, Jimmy Adamson, Alex Anderson,
Tom Mcallister and Jimmy Patrick.

Lynemouth members of the St John Ambulance Brigade in 1959, pictured
outside the pit baths and displaying the Colliery First Aid Senior Team
trophies. They were Area and Divisional Champions in 1957,
1958 and 1959 and runners up in the National Finals in 1959.

A group of mining trainees in 1950 at Ashington Colliery, with their instructor standing in the centre. Prior to the Second World War there was little training for young boys on the surface or underground.

assessment bandaging, moving and carrying casualties. I enjoyed that and when I went back to Lynemouth Colliery continued there and got lots of encouragement from a big squad of St John Ambulance members which included Bill Clancey, Jim Jeffries and Horace Madeley, and I eventually made the First Aid Team.'

Chapter 10

Strikes and Lockouts – For Pay, Pride and a Future

The Miners' Strike of 1926 began on 3rd May and lasted almost seven months. It started when coal owners demanded a 13% reduction in miners' wages and an extra hour on their working day. This was accompanied by a proposal that the minimum wage agreed in 1924 should be abolished. Initially, the miners were supported by the TUC, but after only nine days were left to fight alone when all other trade unions involved capitulated and went back to work.

It would be 1972 before any official national strike action was seen again in British pits. This stemmed from the pit closures of

The seven month strike of 1926 took its toll on the miners' children.
Here at Linton School they are tended by the local community who supplied
what food they had to give the young ones at least one good meal a day.
However, many were malnourished as the Chief Medical Officer
stated after the strike was over.

the 1960s, which included 35 collieries in Northumberland, and the government's energy policy which favoured the oil and gas industries. In 1970 the miners were way down the league in the wages list and when MPs were given a 38% wage increase in 1971 but there was no settlement at all between the NUM and the NCB, an all-out strike was called from 9th January 1972. The response from the public was overwhelming – it was not just about the miners, it was a fight for every low-paid worker in the country.

My first day picketing was when my dad dropped me off at the pit gates at Ellington Colliery. I came face to face with the two NUM diehards, old Denis Murphy and Sam Scott, who stood like banty cocks in front of a large group of boisterous workmates.

'Where yi gan?' Denis asked.

'Aa'm here ti picket, man.'

The strike of 1926 delayed the building of houses in Lynemouth village and the sinking of the pit, which started from 1927 and initially linked with Woodhorn when producing coal from 1934.

Where seams outcropped near the surface then miners would take advantage of the coal during periods of strike action. Here is a fine postcard of Amble and Radcliffe men winning coal for their home fires in 1926.

The magnificent miners' Area Offices, Burt Hall in Newcastle named after Thomas Burt the much respected Northumberland miners' leader and MP.

The men who served on the Executive Committee consisted of miners from pits all around the Northumberland Coalfield. Pictured are members of the 1963 Committee.

Then Sammy piped up, 'Phew, that's great. For a minute Aa thought you were trying ti cross the line. Di' yi hear that lads, another man for the picket line,' and everybody laughed and cheered.

The atmosphere was electric and the strike support almost one hundred per cent. The Deputies worked through the strike as safety cover, but still suffered a rough ride through the picket lines. Lads I had grown up and worked with, some were now Deputies and after they arrived in buses they were escorted by police and had to run a gauntlet of abuse. Mostly it was a push and shove affair, but there were a few nasty incidents.

Stan Elliot was an Ellington union man at this time:

'I picketed at many sites where coal could get through: Stobswood brickyard, Widdrington opencast site and

Miners on the march during the 1972 strike led by their NUM officials, with each lodge banner held high as they parade peacefully through the main street of Ashington.

Druridge Bay to name a few, but there was little movement of coal. It was nothing like as violent as '84. We could pick coal from the old heaps at Widdrington and no one bothered us. One night a few of us had coal in the back of a car when going home and the boot flew open. A policeman stopped and asked what was wrong. "The boot won't stay shut," we said, and he helped us close it and never yet asked what we were carrying.

'One day picketing at Ellington, I persuaded the Deputies to get back on their bus and not go in to work. That was great news as I needed to get to Ashington and so I hitched a lift with them. Well, they were arguing among themselves and then decided they would go into work. The bus stopped at Lynemouth and then turned around and headed back to the pit with me inside. So I had to ring the bell to get off. No way was I going back to the pit with them. Me, Bob Waddell and big George Johnstone went around to the various coal sites with a caravanette providing soup and sandwiches for the lads. One day Joe Grieve cooked a 30lb turkey at Ashington Hirst Welfare and we had the job of dishing it out to the picket sites. We kept some back for us as we had had nowt to eat and were picketing at Lynemouth that night. There was a big fire going in a brazier and we got chips and mushy peas from the fish shop. Tommy Chester turned up just as we were finishing off and called us worse than muck 'cos he had been away picketing and knew nowt about the turkey.

'Henry Cleverley picketed at his home pit of Ellington and down the road at nearby Lynemouth and spent spare time hunting for coal to fuel the home fire. Lots of the lads scavenged for sea coal off the beach and me and others picked at the seams that out-cropped along Lyne Bay. We filled sacks of coal and pulled them up the cliffs by ropes. Aa was working as hard as Aa was at the pit just to get some coal for the fire and a bit for the greenhouse.'

The 1974 miners' strike is best remembered as the one that brought about the three-day week and the downfall of Ted Heath's Conservative government. It all started with disagreement over wage demands. An overtime ban was implemented in November 1973 and then on 1st January 1974 the Tory government announced the country would work a three-day week.

Further discussions between the NUM and the Coal Board came to nothing. The NUM balloted its members who gave an 81% vote in favour of strike action to begin on 10th February 1974. The government responded by calling a general election on 28th February. The miners' wage claim was referred to the Wages Board who stated the miners were due 8% more than they had been offered. It was the end of the road for Heath and his government when they lost the election, and the incoming Labour Party had little option but to settle the miners' claim. As in the 1972 dispute, the strike was solid throughout the country, no movement of unnecessary coal so little need for picketing.

The 1984 miners' strike was the longest and most brutal encounter since the 1926 stoppage. This was not just about pay or conditions, this time around the pit unions were fighting for their livelihood against a full-scale closure of British pits. Plans had been laid down as far back as 1979 with the coming of the Tory government with legislation passed on picketing rules and social benefit payments for strikers. The men stood to lose the first £14 of any income. Working wives lost income support entitlement and single miners received nothing at all.

Jack Tubby gives his version of how things happened:

'Prior to the start of the Ellington Lockout on 22nd February 1984, Ian Macgregor, Chairman of the National Coal Board, paid a visit to Ellington Colliery. He was to go underground and visit the Third North Main Development, but the men decided not to come to work that day. We knew the visit would not be easy because of the overtime ban.

'A large gathering of men congregated outside the main

111

A large, well-schooled contingent of police face pickets at the Ashington Central Workshops in 1984.

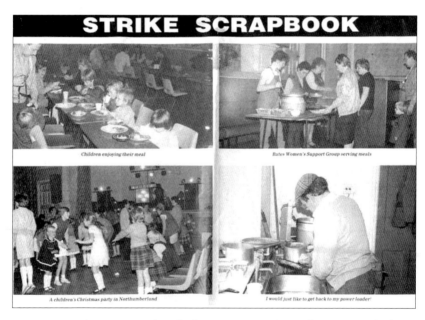

Women support groups played a huge part in assisting their men in the big strike of '84 through fundraising, organising functions and feeding the children, as well as attending morale boosting rallies.

office demanding to see the Chairman. He did offer to speak to union delegates, but this was declined. Sadly, trouble exploded as he came out into the car park, as the crowd surged forward and he ended up on his back. The Chairman of the Ellington Branch of the NUM apologised for what happened. The majority of the workforce were very humble about the incident and I am convinced that the situation was fuelled by external forces and the fact that Macgregor was brought in by the back door. I lost no time in expressing my disgust at the treatment of the Chairman who, after all, was visiting one of the most successful collieries in the country.'

When the closure of Cortonwood Colliery in Yorkshire was announced on 1st March 1984 on economic grounds, this was the catalyst that fuelled the bitter year-long confrontation. Northumberland miners joined the strike after a meeting of its five remaining collieries and Lynemouth Washery Plant at North Seaton on 14th March. A ballot was held the following day. The result was in favour of strike action by a small majority. However, pickets from around the country were turning up at pit gates everywhere and no pitman true to his union ever crosses a picket line.

Northumberland miners had taken up the challenge again just as they had in the disputes of the Victorian era and the stoppages of the early part of the 20th century.

Ellington NUM committee man Jim Sawyer looks back on the dramatic, dark days of March 1984:

'All the lads had for pay was next ti nowt. We paid out a fiver a week from the hardship fund and even paid out the cash once a fortnight to save petrol costs. We then went out on the streets and everywhere with pails to collect money. Most folks were sympathetic towards the strike, but we got abuse from a few people, things like, "Get back ti work, yi buggers, ya costin' the country millions", or

"Ya nay better than beggars – yi get paid well, divvint ya."
This was hard to swallow when we were fighting for our
future. If they had known how the pits would be closed
after the strike they might have taken a different view.'

Alan Stewart of Bedlington was Ellington Mechanics Secretary at
that time and reflects:

'We had a hard core of pickets swelled at times from other
pits in other coalfields. These were your flying pickets who
would go anywhere they were needed. There was a
women's support group who made up food parcels for us
to deliver to pit lads' houses. We always checked their
coal-house first before dropping a parcel off, as some of
the lads were enticed back to work by the management
about November time. If they had a cree full of coals then
we knew they were working so they got nowt from the
union. Some blokes never picketed at all and some local
communities never supported us like they should. As time
went by and some of the lads went back, the frustration
grew as nobody had much to get by on.

'Yes, there were incidents with the police and once I
remember twelve of the lads were to go to Bedlington
Court to answer charges on minor offences. I drove them
there, where they were held in a newly-decorated detention
room. That was not the thing to do and naturally after
they left there was strike support graffiti all over the place.
One lad who was up for breaking a window at the pit by
throwing a stone was asked by the Magistrate, "Did you
throw the stone?" and the lad denied he had. "Come,
come, now. How far were you away from the incident at
the time, a yard or maybe two, or more?" "Well sir," came
the reply, "all I can tell you is that I was more than a
stone's throw away at the time it happened." Pit humour
at its best in times of adversity, eh?

'As the strike wore on there was real hardship, some lads

had big mortgages to pay and bairns to feed. We did have great support from the French and Russian trade unions. I was also part of a delegation that went to Scandinavia to meet trade union reps and drum up support for the strike. When men were filtering back to work, that's when the trouble really started and the scabs were bussed in and supported by the police. The bus drivers wore ski masks to cover their identity and there were grilles on the windows.

'There was once when I was picketing in Lancashire and based at Preston with five other union delegates that a funny incident occurred. After picketing all day on a cold windswept coal site we were as black as the ace of spades. We were staying at the Trade Union Hall, sleeping on the floor with no showers or baths. Having little money, we couldn't afford towels or swimming trunks so we went to the local swimming baths and jumped in fully clothed. That night we just had enough for a pint of beer and went to a bar called Hollywood. Little did we know it was a gay bar and the DJ was a bloke dressed as a woman who sarcastically introduced us as North East strikers. We were led to a sitting end of the bar and soon had visitors, 'cos unknown to us this is where you sat if you wanted a boyfriend! However, the manager took pity on us and provided us with a free drink and tabs.

'We looked for any means of support and one farmer named Hector up at Bamburgh did a deal with the union and left a whole field of tetties for us to dig up. The field was full of us using forks, spades, shovels and anything we could lay our hands on to get the tetties out before they rotted.'

Ian Lavery was an indentured mining craft apprentice at Ellington Colliery when the strike broke out and looks back on events:

'I could have worked during the strike – although an NUM member we indentured apprentices had an agreement from the union to continue our training and studies. But I was becoming disillusioned with management's attitude and joined my father and brothers on the picket lines. This proved to be a major decision and changed my whole outlook and direction in life.

'Yes, I was active on the picket lines and one of the first in the county to be arrested when picketing at Blyth power station when pit folk were fighting for their very existence. Like lots of other lads, I was arrested a number of times and beaten and bloodied. I have no regrets whatsoever doing what I did and proud now to have worked my way up to be the National President of the NUM. This position I could never have held if I had worked through the strike. How could I have looked men in the eye when representing them and delivering stirring speeches if I had not supported them in their hour of need?'

Dave Brown, an Ellington face-worker, picketed as far away as Lancashire:

'All I had was a fiver for the day and we lodged in a school campus. My wife was earning little at a corner shop in Ashington and things got really hard as time went on.

'Breakaway unions supported by the Tories and police tactics at the picket lines did nothing to foster relations between the two parties. The moderate Nottingham, Lancs and South Derbyshire miners were a thorn in the side for the NUM cause. Taunts from the police who would tell striking miners to keep up the dispute as they were making lots of cash out of their overtime and were looking to book a holiday overseas with the proceeds certainly fuelled the situation, which led to many ugly confrontations.'

Sam Scott, NUM General Secretary at the time of the strike, was scathing in his comments in 1984 concerning the return to work of the NACODS union:

STRIKES AND LOCKOUTS – FOR PAY, PRIDE AND A FUTURE

'They chose to ignore and cross the picket lines, quoting safety in the pits as their excuse when the Coal Board had already refused the NUM offer of safety assistance during the holiday period. They either wish to break the strike or their love for money is the deciding factor in the action taken. To the men who have returned to work in armoured buses with police protection, I say we are moving to a Police State under the direction of the most anti-Union government this country has ever seen.'

Women's Support Groups sprang up throughout the country and Northumberland women were to the fore, giving their striking men and their families as much assistance as possible. Newcastle West End and Brenkley Wives Support Groups were among the first, as Sheila Graham of Westerhope, wife of Brenkley miner George Graham, explained in her book *Their Lesson Our Inspiration*, a full-blown account of the group's involvement which she penned after the strike.

Sheila tells it as it was in the year-long dispute:

'I supported the Newcastle West End Support Group who gave assistance to mining families and donated readily to the Brenkley cause. Then a few of us Brenkley women decided to form our own local group to raise funds and donate all the proceeds to the Brenkley Miners Hardship Fund. Everything possible was done to raise money and we attended meetings and rallies to drum up support for our cause. There were some sad times but some good ones too and we never lost hope that things would come good.

'One occasion that warmed my heart was in December when a contingent of six juggernaut lorries filled with Christmas presents for our bairns arrived in the country. These had been donated by our trade union friends abroad in France, Belgium, Denmark and Germany and a Norwegian charity gave £35,000 in Co-op vouchers for

The Bates Colliery Branch NUM Committee who served during the '84 strike pictured at their Blyth headquarters. The only working pit in the Blyth Valley area survived many threats of closure during the latter part of its life, but was finally closed down in 1986 with the loss of 1,700 jobs.

our youngsters' Christmas presents. When van loads arrived at our house we filled the two garages and the house, then delivered the presents to families around the area. It was like being Santa Claus and I went to bed that night tired but very happy now knowing our members would at least have peace of mind, having presents for their bairns at Christmas. Throughout the strike Blyth Valley issued £4 food vouchers weekly for deprived families, as did Wansbeck Council and North Tyneside. The County Council issued £188,000 of vouchers and Tyne & Wear set up a hardship fund. Newcastle City Council gave nothing, just as their forebears had done in 1926, until we petitioned them and after eight months into the strike they came on board.'

As the strike wore on into November with Christmas approaching and no signs of a settlement, men began to drift back to work, lured by management who sent out letters asking them to return.

*The strike was almost over by February 1985, with over 1,000 men
back to work in the county's pits and still a police presence
at the entrance to Ellington Colliery.*

There were ugly, violent scenes on the picket lines. At Whittle Colliery there was an incident that showed police tactics at that time. One of the diehard active picketers named 'Psycho' had a short fuse and caused the police lots of 'aggro'. So they called in a monster of a policeman who looked a lot like Jaws in the Bond films. Jaws goaded Psycho into head butting him at the picket line but it was Psycho who suffered and was arrested for his troubles.

The atmosphere was electric with many striking miners incensed by blacklegs returning to work. It was a time when many were in severe difficulties over mortgage payments. Cars were discarded and sold on to make ends meet, and marriages broke up or were strained to the limit. Any luxuries were abandoned, and only the most basic food served at the table. Little wonder whole families were split asunder. Northumberland was the last county in the UK to have men cross a picket line and return to work, when two miners returned under police protection at Ellington Colliery in August 1984.

The men who returned to work received wages and concessionary fuel and by January 1985 over 1,500 Northumberland miners were back at work. Still the confrontations continued away from the protection of the pit battle buses. In pubs and clubs, houses, back streets and public places, war was waged. Families and life-long friends were torn apart by a rift that in some cases has never been healed.

Many of the NUM activists who were convicted of offences during the strike were dismissed by the Coal Board when it was over and the reason stated for their sacking was always 'gross misconduct'. Most were cleared of any wrongdoing by the law courts yet 17 miners from the Northumberland pits lost their livelihood, with only a small minority being reinstated later. Of the more serious offences prison terms of up to two and a half years were handed out to the men concerned.

Chapter 11

The End Game

he NUM's concerns over pit closures were well founded. After
the strike it was announced that Brenkley, which had serious
geological problems, would close in December 1985. At that
time the pit employed 635 men. Ashington followed in October
1986 and Bates Colliery at Blyth with a workforce of 1,700.
When Whittle Colliery closed in 1987 with the loss of 587 jobs
this only left Ellington Colliery, known as the Big E, to carry on
deep shaft mining in the county.

Opencast coal mining had been introduced to Northumberland
in 1943 to provide extra fuel to support the war effort. With the
closing down of the Big E at Ellington in 2005, opencast is now
the only method of coal mining operating in the county. Recent
media reports brought news of a proposed state-of-the-art method
of mining coal which would consist of ensuring no harmful
emissions of greenhouse gases would pollute the atmosphere, all
residue being contained underground.

All the men I spoke to were devastated at the closing of
Ellington Colliery after an inrush of water from old workings. Kit
Miller told me:

'There were plenty wet pits in the county and no wetter pit
than Whittle in terms of the amount of millions of gallons

In February 1994 it was the end of a life spent in the bowels of the earth when the last pit ponies to work in a British deep shaft mine were brought to bank at Ellington Colliery. Tom, Flax, Carl and Alan are shown with their handlers after emerging from the darkness.

of water we pumped out. If the will had been there, Ellington would never have closed.'

Stan Elliot said:

'It's not so bad for those men coming up to retirement and not needing work, but for all the young lads and the men up to their fifties, there are no jobs around here. What can they do? Move away, retrain for what or take a low paid job? I reckon that's just about it when there was no need to close the pit.'

Ian Lavery, NUM President, was at the meeting with Ellington

A sad day for everyone as miners head for the pit baths for the last time when Ashington Colliery closed in 1986. Mining at this site began in 1866 and by the 1920s the workforce had risen to over 5,000 and Ashington became known as the biggest mining village in the world. In the 1950s the flagship of Ashington Coal Company was still employing over 4,000 staff.

union officials and colliery owners UK Coal when the notice of closure was given:

'We were all stunned when Gerry Spindler, the company's Chief Executive, opened the meeting with the phrase, "Gentlemen, the pit is closed." It took less than three seconds and was an appalling decision after the heroic efforts of the men to control the flooding and real progress was being made.'

John Richardson of Bedlington said:

'It's hard to take after all the effort and miles of pipes we used to control the water, the company said money was no

East Chevington suffered the fate of many of the old colliery villages by being demolished in 1980 to make way for opencast mining. For the people, many of whom had lived here all their lives, it meant being shipped out to the new township of Hadston and surrounding villages. Seen here are the abandoned officials' houses in the foreground and the workmen's houses in the rear awaiting demolition.

object but that is not the case.' Michael Moyle of Amble, who was 38 and had been at Ellington for 14 years, agreed: 'We worked our socks off and pumped out 80 yards of roadway and that shocked the company but still they closed us down.'

Ken Cappelman of Lynemouth, married with two grown-up children, was 56 when the pit closed and said:

'It was a kick in the teeth for everyone, but I felt sorry for the lads aged around the 30 and 40 mark with families and mortgages and with no jobs around the area.'

As Jack Tubby, who managed the Combine in the 1980s, told me:

'Ellington Combine was the biggest undersea mine in the world, breaking production records year after year during the 1980s and once employed 3,000 men above and below ground. There are many who believe it didn't have to end that way, the water problem could have been overcome as it was in various ways in many other mines in the county. It's a sad end to a pit that still boasts vast reserves under the North Sea. A way of life for the miners and their families for generations has disappeared forever.'

February 17th 2006 was the date that signalled the end of deep shaft coal mining in the county. The Ellington Colliery Betty pit shaft tower which dominated the village for almost a century was brought crashing to the ground in just a few seconds, watched by a crowd of locals and media crew.

Gone, but not forgotten. Although the deep mines in the county have now all been abandoned and most of the shafts sealed, our heritage is still maintained and remembered today.

Index